Charles Darwin

THE ORIGIN OF SPECIES

Revised Edition

Abridged and Introduced by

PHILIP APPLEMAN

W • W • NORTON & COMPANY • *New York* • *London*

The text of this book is composed in Fairfield Medium
with the display set in Bernhard Modern.
Composition by PennSet, Inc.
Manufacturing by the Maple-Vail Book Manufacturing Group.
Book design by Antonina Krass.
Production manager: Benjamin Reynolds.

Library of Congress Cataloging-in-Publication Data

ISBN 0-393-97867-2 (pbk.)

W. W. Norton & Company, Inc., 500 Fifth Avenue, New York, N.Y. 10110
www.wwnorton.com

W. W. Norton & Company Ltd., Castle House, 75/76 Wells Street,
London W1T 3QT

1 2 3 4 5 6 7 8 9 0

Contents

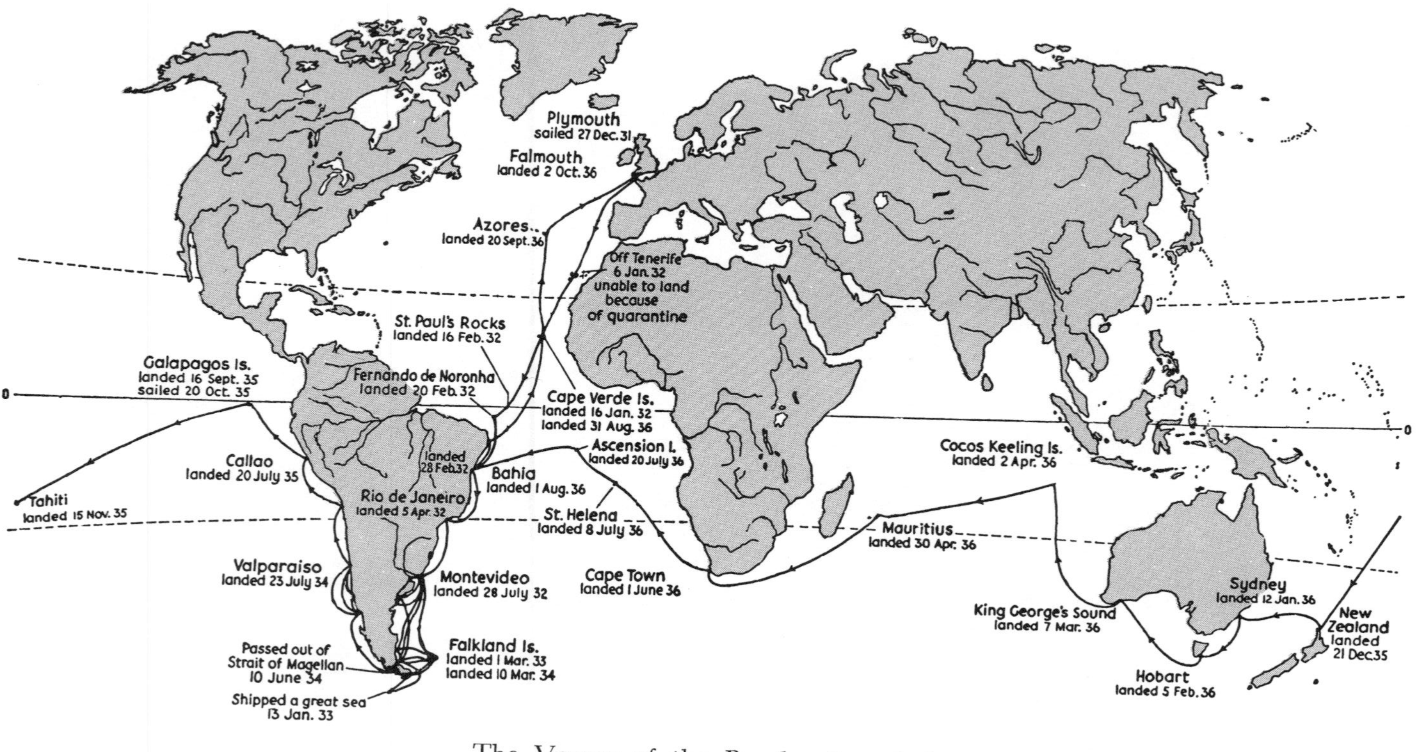

The Voyage of the *Beagle* 1831–1836

Introduction

Born on February 12, 1809, Charles Darwin grew up in the comfort and security of the well-to-do Darwin and Wedgwood families. His mother was a Wedgwood, and he himself was to marry another, his cousin Emma. The son and grandson of prosperous physicians, he tried medical training himself but found the studies dull, and surgery (before anesthesia) too ghastly even to watch. So he followed the advice of his formidable father (six feet, 2 inches; 336 pounds; domineering in temperament) and went up to Cambridge to study for the ministry.

Darwin, however, was less interested in theology than in entomology; since childhood he had taken great pleasure in the popular Victorian hobby of collecting and studying beetles. To obtain his degree, he somewhat impatiently went through the requisite three years of courses in classics, mathematics, and philosophy. At the same time, though, he was also able to study not only insects but the natural sciences in general, with learned professors like the botanist John Stevens Henslow, the geologist Adam Sedgwick, and the philosopher of science William Whewell. He also read John Herschel's impressive new book on scientific method and William Paley's arguments for design in nature. After a substantial education in a broad spectrum of scientific work, Darwin took his B.A. at Cambridge in 1831.

Then a remarkable turn of events saved him from a country parsonage. Professor Henslow unexpectedly arranged for Darwin the invitation to join H.M.S. *Beagle* during a long voyage of exploration. The British government—in the wake of the Napoleonic wars and on the brink of a great commercial expansion—had undertaken the ambitious task of mapping the ports and coastlines of the world. British naval ships were being sent to remote destinations, many of them carrying along naturalists, who were expected to make their collections and observations while the mapping expeditions were carried out. Justifying his nomination of so young a man for such a position, Henslow wrote to Darwin that although he was not "a *finished* naturalist," he was already "amply qualified for collecting, observing, and noting anything worthy to be noted in Natural History." So on December 27, 1831, at the age of twenty-two, Darwin left England for what became a five-year journey around the globe; it turned out to be not only a crucial experience for Darwin himself but a passage of consequence for the whole world.

During the voyage of the *Beagle*, Darwin resolutely transformed himself into a "finished naturalist"—an industrious collector, a keen observer, a canny theorist. And he took up the momentous problem that he would grapple with for more than twenty years: what he called the "mystery of mysteries"—the origin of species. In the years after his extraordinary journey on the *Beagle*, however, Darwin's adventures were mostly intellectual, his life deliberately domestic, primarily because he was chronically ill for much of his later life with a mysterious, undiagnosable ailment. He suffered from heart palpitations and an almost daily debilitating nausea, the cause of which is still not certain. So he lived quietly in his country house[1] with his beloved and devoted wife, Emma, who managed their large household of children, servants, and pets. Emma attended her husband assiduously through his chronic sickness, while caring for their ten children; and she sustained him in his work, asking him challenging questions, writing countless letters at his dictation, and helping correct the proofs of his weighty books.

Despite his continuing ill health, Darwin worked hard almost every day, and his industrious life was studded with solid contributions to science: *The Voyage of the Beagle* (1845), *The Origin of Species* (1859), *The Variation of Animals and Plants under Domestication* (1868), *The Descent of Man* (1871), *The Expression of the Emotions in Man and Animals* (1872), *The Formation of Vegetable Mould through the Action of Earthworms* (1881), and so on—as well as an autobiography and a voluminous worldwide correspondence on scientific matters. (See Selected Readings, p. 123.)

There was something paradoxical but eminently admirable about both Darwin's character and his devotion to his task. Intellectually he was a revolutionary, but the gentlest of revolutionaries. He lived the life of a respectable and conventional country gentleman, but he gradually developed some very unconventional ideas and, as he wrote in his autobiography, ultimately rejected the "brutal" religion that threatened to condemn his freethinking father, brother, and best friends to everlasting punishment ("a damnable doctrine"). He became, in T. H. Huxley's new terminology, an agnostic; and like many another Victorian agnostic, he exemplified in his life and work a high-minded benevolence, kindness, and generosity not only to other people but to all creatures. So he continued to write about

1. Down House, about fifteen miles southeast of London, near the village of Downe, in Kent; now a Darwin museum open to the public.

the "grandeur" of "beautiful" and "wonderful" forms of life, and of humanity's high "destiny" in the future.

When he died in 1882 at age seventy-three, the man whose sacrilegious ideas had once been publicly assailed by a multitude of critics was accorded the rare national honor of burial in Westminster Abbey, a few feet from the grave of the other eminence among British scientists, Isaac Newton.

2

It is not easy, a century and a half later, to enter fully into the mind-set of the early Victorians, in which both religion and science presupposed the glory of God to be manifested in nature. Nor is it easy to comprehend the intensity of their commitment to the fixity of species, or to imagine the intellectual and emotional upheaval that Darwin's work would cause. And it is difficult to give sufficient credit to Darwin's boldness and originality, unless it is kept in mind that he had boarded H.M.S. *Beagle* young, as yet unseasoned in science, and still a believer in Genesis.

We need to picture him then, just out of Cambridge, carrying in his small shipboard library the first volume of Charles Lyell's revisionary new work, the *Principles of Geology*, but warned by his respected professor Henslow against its novel uniformitarian ideas. He also carried in his mental baggage the pious lessons of William Paley's *Natural Theology*, which he had studied at Cambridge so carefully that he later wrote in his autobiography, "I could almost formerly have said it by heart." And it was Paley more than anyone else who had already persuaded a generation of readers that in the Deity's neatly contructed universe, "the marks of [God's] *design* are too strong to be gotten over."

Darwin also kept in mind the ideas and opinions of the Cambridge professor William Whewell—"next to Sir J. Mackintosh," Darwin wrote, "the best converser on grave subjects to whom I ever listened." By an ironic trick of history, it was during the five years of Darwin's *Beagle* voyage that the British citadel of science, the Royal Society, was administering the publication of the Bridgewater Treatises, a series of books by notable scientists and moralists, including Whewell, who were commissioned specifically to demonstrate "the power, wisdom, and goodness of God as manifested in the Creation." So, while Darwin was in South America painstakingly examining plants and animals—assembling the physical evidence that would one day support his radical new theory—Whewell, in his Cambridge study, was writing for his treatise:

> If there be, in the administration of the universe, intelligence and benevolence, superintendence and foresight, grounds for love and hope, such qualities may be expected to appear in the . . . fundamental regulations by which the course of nature is . . . made to be what it is.

It is an awesome distance from that kind of thinking to Darwin's in *The Origin of Species*, which, a quarter of a century later, would turn Whewell's divinely planned, benevolent world topsy-turvy:

> Thus, from the war of nature, from famine and death, the most exalted object which we are capable of conceiving, namely, the production of the higher animals, directly follows.

Charles Darwin—cautious, skeptical, compulsively industrious, distrustful of his own talents, and never daring to suspect himself of the genius we now acknowledge—did not make that awesome voyage in a day.

As a modest young scientist, Darwin was understandably reluctant to reveal his revolutionary theory of the transmutation of species and thus set himself single-handedly against the massive forces of conventional scientific and religious opinion, both of which were committed to the ancient and sanctified belief in the fixity of species. He knew about Lamarck's bitter experience: Lamarck had tried to challenge that conventional opinion with an unconvincing evolutionary hypothesis and had been systematically attacked and ridiculed by virtually the entire scientific establishment. Other scientists, philosophers, and writers (including Darwin's own grandfather, Erasmus Darwin) had also speculated about the transmutation of species, but, like Lamarck's, their work was never taken seriously either; it was too hypothetical or too superficial to threaten in any serious way the established scientific and religious belief in the fixity of species.

Faced with such daunting odds, Darwin wisely avoided publishing anything about evolution, but he continued to build the evidence for his developing theory. Even as a young man, he had been a tenacious empiricist, a tireless gatherer of facts. During the *Beagle* voyage, for five years, from one exotic locale to another, he had explored riverbeds and coral reefs, hiked pampas and climbed mountains; he had recorded stratifications of rocks and soil and examined the earth with lens, compass, clinometer, penknife, blowpipe, and acids; he had discovered fossils of conifers and shellfish, of Megatherium and mastodon. During those five industrious years, he had collected and continuously shipped to England vast numbers of plants, birds, insects, reptiles, and fossils. And during those five early years he had also become a subtle theorist, pondering the causes of the elevation and subsidence of volcanic strata, of the earthquakes he had experienced, and of the formation of coral reefs.

After the *Beagle* voyage, the maturing Darwin, now widely respected as a naturalist, spent the next twenty years in a dogged pursuit of his evolutionary hypothesis—examining the many breeds of domestic pigeons, the skeletons of rabbits, the wings of ducks, the variations in ten thousand specimens of barnacles; keeping notebooks on "transmutation"; and discreetly discussing the "species question" with Charles Lyell and other close scientific friends. Finally, in middle age, he dared to begin to write the ambitious book that he intended to call "Natural Selection."

But even then—after five years of exploration on the *Beagle*; twenty years of patient study, careful observation, and cautious speculation; and then two more years of exhausting work on the steadily growing manuscript of "Natural Selection"—even after all that hard labor and deep thought, Darwin was still reluctant to publish his challenge to the scientific and religious establishment.

Then, on June 18, 1858, everything changed. On that day, Darwin received the momentous letter from Alfred Russel Wallace, describing Wallace's own recent discovery of the principle of natural selection. Darwin immediately wrote to Lyell: "Your words have come true with a vengeance—that I should be forestalled." But his dismay was temporary. Lyell and Joseph Hooker, the botanist, arranged a joint presentation of short papers by both Wallace and Darwin at the Linnean Society in London in July 1858; so their names became permanently linked as codiscoverers of the principle of natural selection. Wallace, however, was always modest about his contribution, because, as he said, compared to Darwin's, it was as two weeks are to twenty years.

After Wallace's unintentional prompting, Darwin realized that he should finish and publish his lengthy manuscript expeditiously; but to do so, he would have to restrict its size. So he condensed the project into what he called an "abstract" of his work, and by March 1859, he had completed *The Origin of Species*. It was published on November 24 of that year. The effects of that publication were immediate, extensive, and profound. Hardly any kind of thought—scientific, philosophical, religious, social, literary, or historical—remained long unchanged by the radical implications of the *Origin*. That extraordinary, continuing, and worldwide transformation is the subject of this volume.

3

There were various reasons for the remarkably swift spread of evolutionary thought after the publication of the *Origin*, and one of the most important was the Victorians' fascination with science in general. Thomas Henry Huxley wrote in 1873, "We are in the midst of a gigantic movement, greater than that which preceded and produced the Reformation." He was referring to the persuasiveness of the scientific method, so clearly burgeoning at the time.

However, for most people, then as now, what was meant by "science" was not the scientific method—a way of organizing data, hypothesizing about them, and generalizing from them—but rather applied science, that is, technology. During the long reign of Queen Victoria, from 1837 until her death in 1901, scientific technology transformed many of the conditions of people's lives. The first railroad was built in England in 1825, when Victoria was a little girl. Before that, the maximum speed of land travel was—for up-to-date Britons as it had been for Caesars and pharaohs—the speed of the horse. But by the time the venerable queen died, almost all of Britain's extensive railway system had been built. "Science" had liberated travel from animal muscle and had begun that acceleration toward inconceivable velocities, which is so characteristic of our own age and is as impressive to us as it was to the Victorians.

Impressive: "science" was *doing* things, making things *work*. While Victoria was still on the throne, transatlantic steamship service began, power-driven machines revolutionized industry, the telegraph and telephone were developed, and many other new inventions were put into use, from the electric light to the automobile. In 1851, eight years before *The Origin of Species*, the Victorians celebrated Progress at the first World's Fair in London—the Great Exhibition of the Industry of All Nations—in the splendid Crystal Palace, where it was abundantly demonstrated that "science" was indeed making things happen; it seemed, to many at the time, both fascinating and dependable.

Unlike philosophy or theology or literature, science compels rational assent. And ever since Copernicus shunted people off from the center of the universe, no scientific discovery had been as staggering as Darwin's. One could not simultaneously accept his evidence and the literal words of Genesis. No reconciliation was possible, T. H. Huxley insisted, "between free thought and traditional authority. One or the other will have to succumb."

What was at stake was no less than a worldview. In the seventeenth century, ecclesiastics had calculated that God created man at 9:00 A.M. on October 23 in the year 4004 B.C.E.—a year still cited in some annotated King James Bibles. And Paley had argued confidently in 1802 that the universe was carefully designed by a provident Creator. However, natural selection functions not by design but by incessant opportunism, and it is an extremely time-consuming process: six thousand years are a mere instant on the evolutionary time scale. Paley's view of the world was familiar, pious, and comforting. Darwin's view of the world was revolutionary, disturbing—and persuasive.

Because it was so persuasive, *evolution* became a watchword for the late Victorians. By the end of the nineteenth century hardly any field of thought remained unchanged by the exciting new concept. Historians began looking at the past as a living, changing organism; legal theorists studied the law as a developing social institution; critics examined the evolution of literary types; anthropologists and sociologists invoked natural selection in their studies of social forms; apologists for the wealthy showed how the poor are the "unfit" and how progress, under the leadership of the "fit," was inevitable; novelists "observed" their characters as they "evolved" in an "empirical" way; and poets invoked an evolutionary life force. In 1800 the word evolution had been used only to signify the development of organisms from their immature to their mature state; but in the half century following the publication of *The Origin of Species*, evolution seemed capable of explaining anything and everything. The titles of ponderous books of the period indicate how religion itself was being reconsidered: *The Evolution of Morality* (1878), *The Evolution of Religion* (1894), *The Evolution of Immortality* (1901), *The Evolution of the Soul* (1904).

4

Science is not a single enterprise but a host of different activities, all having in common a special "way of knowing" that was stimulated by a new emphasis on inductive reasoning in the seventeenth century, and has been constantly examined and amended ever since. Darwin's scientific work not only was a major contribution to knowledge in and of itself but also has been seminal for later biologists. Now, a century and a half after *The Origin of Species*, generations of dedicated scientists have labored to refine and augment Darwin's groundbreaking idea: discovering new evidence, developing new fields of study, changing emphases, and working out the thousand details of what Ashley Montagu called "the gigantic complex jigsaw puzzle that is evolution." Fitting together the pieces of that puzzle has required innumerable contributions from astronomy, biology, chemistry, biochemistry, physics, paleontology, geology, genetics, developmental biology, ecology, anthropology, geochemistry, geophysics, and other disciplines—as the following examples indicate.

- Paleontologists are revealing countless new facts about the human genealogy; about our evolutionary origins in Africa;

and about our prehistoric incursions into Asia, Europe, and the rest of the world.

- At the announcement of the sequencing of the human genome in June 2000, geneticist Jon Seger pointed out that the genome is "evolution laid out for all to see." And Nobel laureate David Baltimore added, "The genome shows that we are all descended from the same humble beginnings and that the connections are written in our genes."
- Molecular biologists continue to inform us authoritatively about the close relationship of humans with other animals, especially, of course, with the other primates—for instance, our more than 98 percent genetic identity with chimpanzees.
- Collaterally, primatologists are generating voluminous new information about the intelligence of the great apes—and even, in the case of chimps, about their culture—which is causing us to rethink our attitudes, and our obligations, to our close biological cousins.
- Taxonomists are now working with the relatively new method of biological classification called cladistics—a method that focuses closely on genealogy and evolutionary "branching order"—which has altered some older classifications and has, among other things, helped sort out our own recent biological forebears.
- Naturalists in the Galapagos have painstakingly observed Darwin's finches, establishing the fact of natural selection operating in real time—that is, in the biologically brief period of a few decades. This is one of many recent studies in which natural selection has been subject to human observation.

As these and other important developments continue to demonstrate, natural selection remains fundamental to all fields of biological research; and biologists universally acknowledge and respect Darwin as the architect of modern biology.

5

Discussions of Darwinism sometimes become analogical rather than analytical, "Darwinistic" rather than Darwinian. Morse Peckham made this useful distinction:

> Darwinism is a scientific theory about the origin of biological species from pre-existent species. . . . Darwinisticism can be an evolutionary metaphysic about the nature of reality and the universe . . . an economic theory, or a moral theory, or an aesthetic theory, or a psychological theory.[2]

2. Morse Peckham, "Darwinism and Darwinisticism," *Victorian Studies* 3 (1959), 32.

Having been for years scrupulously careful and objective in his explorations of scientific matters, Darwin grew more and more willing, as time went by and his interests broadened, to speculate daringly on the basis of what he considered the best available knowledge. When writing *The Origin of Species*, Darwin characteristically exhibited a scientist's proper caution in language like this:

- "We must not overrate the accuracy of . . ."
- "We must be cautious in attempting . . ."
- "Judging from the past, we may safely infer that . . ."
- "We may look with some confidence to . . ."

And so on. As Alfred North Whitehead observed, "Darwin's own writings are for all time a model of refusal to go beyond the direct evidence, and of careful retention of every hypothesis."[3]

Twelve years later, in *The Descent of Man*, a more mature and experienced Darwin continued to exercise that same caution in discussing human bodily structure, embryonic development, and rudimentary organs. However, when he turned his attention to mental and moral qualities, he had less tangible, physical evidence at hand and had to rely more on anecdote and speculation, both of which are, of course, subject to cultural distortion. Darwin could not completely overcome this disadvantage; like everyone else, he was a product of his times, although he was often ahead of them. Having settled down to a conventional life in a rural village, he undertook the responsibilities of a respectable landowner and accepted without any serious question the gender roles of the society that had so generously nurtured him. Consequently, he was sometimes guilty himself of straying from Darwinism into some Darwinisticism or other, including what we are now accustomed to calling a sexist denigration of the intellectual abilities of women. But as Evelleen Richards points out,

> Darwin cannot be personally judged by twentieth-century yardsticks any more than his work can be assessed by twentieth-century standards and concepts. To label him a sexist may be technically correct . . . but is mere rhetoric in the context of a society in which almost everyone . . . held discriminatory views of woman's nature and social role.[4]

But if Darwin strayed innocently into an occasional Darwinisticism, others took that step knowingly and intentionally. Some people saw an opportunity to take over his new concepts and exploit them as self-serving rationalizations. Social Darwinism, the most

3. Alfred North Whitehead, *Science and the Modern World* (New York, 1925), 158.
4. Evelleen Richards, "Darwin and the Descent of Woman" in *The Wider Domain of Evolutionary Thought*, ed. David Oldroyd and Ian Langham (Dordrecht, Holland, 1983), 99.

obvious example of this distortion, is the spurious claim that Darwinian competition in nature constitutes a proper model for the "survival of the fittest" in human society—in which everyone competes to survive, but only the wealthy have proven themselves "fit."

The publication of *The Origin of Species* and the American Civil War were almost coincident, and in the years following that war, the United States, like other Western nations, was industrializing very rapidly. The late nineteenth century became the pre-eminent period of the rugged individualists, the robber barons, the captains of industry, the accumulators of great wealth. However, it was also a harsh period for ordinary working people: a time of sweatshops, union busting, goon squads and strike-breaking massacres, dollar-a-day wages, tenements without sanitation, and widespread malnutrition.

It was not easy, in a "Christian" society, to reconcile such glaring contradictions. Ever since the beginnings of the industrial revolution, the economic establishment had been casting about for rationalizations, for self-justification. Free-market political economy had been just such a sanction, emphasizing as it did the necessity for untrammeled individual enterprise, for "enlightened self-interest," and for the "iron laws" of economics. Andrew Carnegie once said, "There is no more possibility of defeating the operation of these laws than there is of thwarting the laws of nature which determine the humidity of the atmosphere or the revolution of the earth upon its axis."[5]

Oddly enough, there were also religious sanctions for Social Darwinism—for had not the clergy always warned that material things were corrupting to people, and was it not therefore self-evident that the "lower classes" should be kept poor to be kept virtuous? And, paradoxically, could not the virtuous and industrious among the poor expect to be rewarded, even in this world? Ben Franklin had declared:

> He that gets all that he can honestly, and saves all he gets will certainly become rich, if that Being who governs the world, to whom all should look for a blessing on their endeavors, doth not, in His wise providence, otherwise determine.

Finally, as always in the nineteenth century, there was that court of last resort, the sanction of progress, in whose name all contradictions were resolved—or ignored.

Such rationalizations had always been welcomed by the economic establishment; but after 1859, it found in natural selection a universal sanction that gathered into one grand synthesis all of the

5. Andrew Carnegie, *The Empire of Business* (London, 1907), 67.

previous ones and added to them its own scientific prestige. Natural selection and the struggle for existence lent authority to laissez-faire economics: the celebrated Social Darwinist, Herbert Spencer, insisted that the state should refrain from action calculated to interfere with the struggle for existence in the industrial field. Natural selection also reinforced the older religious sanctions: one economist said that the "laws" of natural selection were "merely God's regular methods of expressing his choice and approval."[6] And evolution seemed to make progress inevitable, ending, Spencer predicted, in "the establishment of the greatest perfection and the most complete happiness." In his opinion, progress was "not an accident, but a necessity."[7]

The captains of industry were quick to pick up this Darwinistic vocabulary. John D. Rockefeller told a Sunday-school class that

> The growth of a large business is merely a survival of the fittest. . . . The American Beauty rose can be produced in the splendor and fragrance which bring cheer to the beholder only by sacrificing the early buds which grow up around it. This is not an evil tendency in business. It is merely the working-out of a law of nature and of God.[8]

Richard Tawney once argued that the true character of a social philosophy is defined most clearly by how it regards the misfortunes of those of its members who fall by the wayside. Those who fell by the wayside in late-nineteenth-century America were assured by the conservative apologists that this was their natural lot, that science "proved" that they were the unavoidable by-products of an ultimately beneficent struggle for existence, without which there could be no progress.

However, some people would not accept such a callous doctrine, and questioned whether the implications of evolution were really as somberly deterministic as was being alleged. As early as the 1870s, some naturalists had been investigating instances of cooperation, in addition to competition, in the natural world. In 1902 Peter Kropotkin published his book *Mutual Aid*, revealing that he had

> failed to find—although I was eagerly looking for it—that bitter struggle for the means of existence, *among animals belonging to the same species*, which was considered by most Darwinists (though not always by Darwin himself) as the dominant characteristic of the struggle for life, and the main factor of evo-

6. Thomas Nixon Carver, quoted in Richard Hofstadter, *Social Darwinism in American Thought* (Boston, 1955), 40.
7. Hofstadter, *Social Darwinism*, 40.
8. Quoted in J. Ghent, *Our Benevolent Feudalism* (New York, 1902), 29.

> lution. . . . On the other hand, wherever I saw animal life in abundance . . . I saw Mutual Aid and Mutual Support carried on to an extent which made me suspect in it a feature of the greatest importance for the maintenance of life, the preservation of each species, and its further evolution.

Thomas Henry Huxley had already made the famous distinction, in a notable lecture of 1893, that

> social progress means a checking of the cosmic process at every step and the substitution for it of another, which may be called the ethical process; the end of which is not the survival of those who may happen to be the fittest . . . but of those who are ethically the best.

By the end of the nineteenth century, then, some naturalists were broadening the evolutionary view of life to show that nature was not always, or not simply, competitive. And social and ethical thinkers were attempting to demonstrate that even though fierce competition does exist in nature, it does not necessarily follow that it is a proper pattern for human behavior.[9]

The twentieth century inherited these controversial nineteenth-century issues, and they are still being debated, albeit in altered contexts. Scientists now discuss the extent of genetic influences on both altruistic and selfish behavior. They also consider the role that gender has played in the development of the various human cultures. And they argue about the extension of biological principles, not only into the study of modern psychology and medicine but also into the social sciences, and potentially even into the humanities.

6

A nineteenth-century adage proposed that it is the fate of all great scientific discoveries to pass through three stages. In the first stage, people say, "It's absurd"; in the second, "It's contrary to the Bible"; and in the third, "Oh, we've known *that* all along." Evolution passed through all three stages so rapidly that in the last edition of the *Origin* in Darwin's lifetime (1872), he could write:

> As a record of a former state of things, I have retained in the foregoing paragraphs, and elsewhere, several sentences which

9. Ironically, communism and national socialism, like capitalism, had also been searching for "scientific" sanctions in Darwinism. Karl Marx read into *The Origin of Species* his own preoccupation with class struggle and historical patterns of social change. Later on, Adolf Hitler, too, rationalized the ruthless Nazi social plans as "Darwinian." All three examples—capitalist, communist, national-socialist—demonstrate how seductive Darwinistic pretentions can be, and (by their mutual contradictions and untrustworthy analogies) how unreliable they are. [Editor]

> imply that naturalists believe in the separate creation of species, and I have been much censured for having thus expressed myself. But undoubtedly this was the general belief when the first edition of the present work appeared. I formerly spoke to very many naturalists on the subject of evolution, and never once met with any sympathetic agreement. . . . Now things are wholly changed, and almost every naturalist admits the great principle of evolution.

Scientific revolutions depend on, among other things, the insight and the determination—and sometimes the aggressiveness—of their protagonists. Darwin had the insight to discover natural selection, and he had the determination to fill hundreds of pages with minutely observed facts and close reasoning in support of it. Friends, when necessary, supplied the aggressiveness—and the world was changed: converted, in a very few years, from an almost total belief in the fixity of species to a widespread understanding of the transmutation of species.[1]

That in itself was a remarkable conversion, but the Darwinian revolution did not stop there; it also required a basic change in thinking about all ideas, all phenomena. To the conventionally religious, the most threatening thing about Darwinism was the implication that nothing was sacrosanct: evolution was becoming not only the science of sciences but, even more disturbing, the philosophy of philosophies.

For those hostile to such change, the impact was shocking, and counterattacks from the faithful were swift and fierce. In the pages of sectarian periodicals, where alleged scientific experts could flourish in protective anonymity, priests, parsons, and bishops defended not only their own faith but also "true Baconian induction." A truculent contributor to the *Catholic World* wrote:

> The theory of evolution has no scientific character, is irreconcilable with the conclusions of natural history, and has no ground to stand upon except the worn-out fallacies of a perverted logic.[2]

The aging Anglican sage Dr. Pusey sternly rejected Darwinism:

> Never probably was any system built upon so many 'perhaps,' 'probably,' 'possibly,' 'it may be,' 'it seems to be' . . . as that

1. Although the *principle* of evolution was, as Darwin wrote, widely accepted in his own lifetime, scientists continued to debate the *mechanism* of evolution for decades—disputes that could not be resolved until the rediscovery of Mendelian genetics in 1900 and the mathematical work of Ronald Fisher in 1930, which led quickly to the modern neo-Darwinian synthesis and the confirmation of natural selection as the primary agent in evolutionary change. [Editor]
2. "Dr. Draper and Evolution," *Catholic World* 26 (1878), 775.

> mythological account of the origin of all which has life, and, at last of ourselves, which is now being everywhere or widely acknowledged by unscientific minds as if it were axiomatic truth.[3]

Pope Pius IX, writing to a French anti-Darwinian author, was also thoroughly contemptuous:

> We have received with pleasure, dear son, the work in which you refute so well the aberrations of Darwinism. A system which is repudiated by history, by the traditions of all peoples, by exact science, by the observation of facts, and even by reason itself, would seem to have no need at all of refutation . . . [But] such dreamings, absurd as they are, since they wear the mask of science, [must] be refuted by true science.[4]

It need hardly be pointed out that the concern of theologians for "true science" was not wholly disinterested. They worried because they saw, perhaps more clearly than others, the philosophical implications of post-Darwinian thought. It was not just that Darwin had undermined the Book of Genesis, or even that he had given scientific authority to the nineteenth-century affinity for endless continuities rather than eternal verities, or that the evolutionary orientation stressed context and complexity—although all such ideas threatened established religion. The most dangerous idea of all was that Darwin's universe operated not by design but by natural selection, a self-regulating mechanism. "Paley's divine watchmaker was unemployed," Gavin de Beer wrote in 1964, "because the wonderful property of organisms is that they make and mar themselves." Natural selection pictured the world in a constant process of change, but without any apparent prior intention of going anywhere in particular or of becoming anything in particular.

That was a devastating proposition for conventional believers—more so, perhaps, than the Copernican theory had been, because it struck even closer to home. Science, with all its impressiveness of fact and achievement, was moving in on the theologians, *taking over*—constantly enlarging the domain of fact and consequently reducing the domain of speculation. Given objective knowledge, most people tend (it was already clear) to give up the guessing games of ignorance—such as conjecture about which of the "humors" is overbalanced in a person whose blood pressure is abnormal or whether or not Adam and Eve had navels or what influence the planets have on our destinies. So it *made a difference* that humans had not only

3. Edward Bouverie Pusey, *Un-Science, Not Science, Adverse to Faith* (London, 1879), 32.
4. Quoted in Constantin James, *L'Hypnotism expliqué et Mes Entretiens avec S.M. l'Empéreur Don Pedro sur le Darwinisme* (Paris, 1888), 84–85.

evolved but had evolved by natural selection rather than by an imaginary vital force or hypothetical cosmic urge. Darwinism was uncompromisingly nonteleological and nonvitalist, and that basic fact was anathema to theologians. Of course, theologians had always seen human beings as a part of nature, but usually as a much grander part: as the crowning achievement of God's universe, only a little below the angels. It was the Darwinian challenge to that lofty assumption that caused tremors among the reverential and the orthodox, who then repeatedly denounced the new heresy.

But with the passing of decades, direct opposition to Darwin gradually made way for accommodation. The opposition was still active in 1909, when the first edition of the *Catholic Encyclopedia* said of natural selection that

> As a theory it is scientifically inadequate. . . . The third signification of the term *Darwinism* arose from the application of the theory of selection to man, which is likewise impossible of acceptance.

However, by 1967, the *New Catholic Encyclopedia* was looking at these things more realistically:

> Today, with a much better understanding of both the theological sources of the Judeo-Christian revelation and of evolutionary theory, the compatibility of God's creative and directive action is more easily comprehended. . . . The solution of the basic difficulties [with Darwinism] was soon found to lie in Biblical research and scholarship and not in the rejection of the new theory. In his encyclical *Humani generis* [1950], Pope Pius [asserted] that general evolution, even of the body of man (and woman) should be professionally studied by both anthropologist and theologian.

More recently, there has been a substantial and continuing movement of theological opinion toward the acceptance of evolution, not only by Catholics but by Jewish groups and by mainstream Protestant denominations, such as Presbyterians, Lutherans, Episcopalians, and others—despite the invention of that flagrant oxymoron "scientific creationism," which is being promoted by a vocal minority of fundamentalist Protestants.

Although the arguments of the creationists are intellectually frivolous, they are being promoted with evangelical fervor and often have serious political and educational consequences. One concrete way of illustrating this point is to recount my own personal experiences in learning about Darwin and evolution.

In the preface to one of my books of poetry, *Darwin's Ark*, I noted some ironic personal coincidences. For instance, I happened to be conceived in the same month in 1925 that John Thomas Scopes was indicted for the crime of teaching evolution to the schoolchildren of Dayton, Tennessee. (Scopes was targeted by a prohibitory law instigated by religious fundamentalists.) In due course, I was born in the same month that the legislature of the state of Mississippi, also under pressure from fundamentalists, passed an anti-evolution law similar to the one in Tennessee. Also in the year of my birth, the famous evangelist Aimee Semple McPherson, concocting an alibi for an extended rendezvous with her lover, claimed to have been kidnapped by "gamblers, dope peddlers, and *evolutionists.*" Then, a year after my birth, the American Anti-Evolution Association was founded, declaring itself open to all citizens except "Negroes, Atheists, Infidels, Agnostics, *Evolutionists*, and habitual drunkards."

By the time I started school, the fundamentalists, continuing their nonstop crusade against science, had caused antievolutionary laws to be passed in Arkansas and in Florida, and were agitating for similar laws in other states. Meanwhile, they also contrived to get high school and college teachers fired for discussing evolution in the classroom. At the same time that my classmates and I were learning to read, textbook publishers, fearing a loss of sales in the Bible Belt, were busy deleting any reference to Darwin or evolution from virtually all public school textbooks in this country. That boycott continued for decades—partly by laws devised by the fundamentalists, partly by collaborating or intimidated school boards, and partly by the publishers' ongoing self-censorship.

So in twelve years of education, including a high school course in biology, I never once heard the name of Darwin or the word evolution. Like millions of other students around the country, my classmates and I graduated from high school totally ignorant of one of the most basic facts of nature: the perpetual functioning of organic evolution. The fundamentalists had actually managed to abolish a natural law from the schools. It was an astonishing feat—the educational equivalent of, say, the Flat Earth Society abolishing gravity.

It would be gratifying if all that were now changed; but unfortunately, the creationist political lobbying is continuing and intensifying. As a result, even today in many American high school biology courses, evolution is still frequently ignored, diluted, or forbidden. The deplorable consequence is that many adults are not able to understand the science of evolution. Worse still, they are uninformed or misinformed about modern science in general—about its reliable methodology and its impressive intellectual achievements. Compounding their misfortune, these educationally deprived people often become the easy victims of assorted charlatans—astrologers, psychics, self-proclaimed prophets, Bermuda Triangle occultists, parapsychologists, UFO visionaries, New Age mystics, numerologists, faith healers, channelers, pyramidologists, fortune tellers—and creationists. Obviously, the scientific and educational community needs to maintain constant vigilance and a vigorous program of public information and public advocacy, if factual knowledge and common sense are to prevail over ignorance and superstition.

I finally did get around to reading Darwin, almost by accident. I had been in the Army Air Corps during World War II, and after the war, started college on the G.I. Bill. Two years later, to help pay tuition and expenses, I joined the Merchant Marine, and periodically spent many months at sea—a good place for earning money while also keeping up with my studies, because there was time for reading in the long hours off-watch. On one extended voyage to the Mediterranean in 1948, I happened to pack in my seabag, along with many other books, the Modern Library Giant edition of *The Origin of Species* and *The Descent of Man*, both in one thick volume: exactly one thousand pages of small print. (I still have that book, a bit dog-eared, its text underlined and annotated with a smudgy ball-point pen.) I was twenty-two—the same age as Charles Darwin when he set out on the *Beagle*—and because of his book, my trip, too, was a voyage of discovery. I remember that experience vividly: the exhilaration I felt, reading Darwin's clear and persuasive observations; the relief at being finally released from a constrained childhood allegiance to the primitive creationist myths of Genesis; the profound satisfaction of understanding the facts of biological evolution and knowing that I was truly and altogether a part of nature.

When I got back to college, I went on reading Darwin, and I have been reading him ever since—*The Voyage of the Beagle; The Expression of the Emotions in Man and Animals*; his books on barnacles, on earthworms, on orchids; and his journals, notebooks, letters, and autobiography—his whole prodigious labor of love. And I have also been reading *about* Darwin ever since, and writing about him: a doctoral dissertation, the Norton Critical Edition, *Darwin*, and articles, lectures, chapters, reviews, novels, and poetry—particularly the novel *Apes and Angels* and the poems in *Darwin's Ark*. So my own adventure with Darwin has ended happily; but the sad fact is that many thousands of young people, even today, are still being denied a sound education in biology as a result of creationist political pressures.

7

Ever since the seventeenth century, literature has been responsive to developments in science—astronomy, physics, and geology; and nineteenth-century writers quickly began to assimilate evolutionary ideas into their work. Tennyson and Browning both wrote about evolution throughout their long careers; and evolution was also cited, discussed, and absorbed in the novels of George Eliot, Bulwer-Lytton, Charles Reade, Mrs. Humphry Ward, George Gissing, and Charles Kingsley, among many others. The writers most influenced by evolution include George Meredith, Thomas Hardy, and A. C. Swinburne, who tended to see the end result of human tribulations as true progress, even in a world that is red in tooth and claw. Meredith built a "philosophy of earth" around this belief, although at times he, like Tennyson, only "faintly trusted the larger hope [that] somehow good will be the final goal of ill." Somewhat later, Samuel Butler, H. G. Wells, George Bernard Shaw, and others also used evolutionary themes in their writing.

Not surprisingly, the philosophical and religious controversies over Darwinism continued to have significant literary consequences in the twentieth century. In one debate, for instance, it was alleged that Darwin's materialistic science eroded human values, and therefore destroyed high tragedy. However, that allegation conveniently ignored the dearth of tragedy in the century before Darwin. In fact, high tragedy at any time is rare; every age does not produce a Sophocles or a Shakespeare.

More to the point, why should it be assumed that a physically comprehensible world is necessarily a morally valueless one? Those who assert that it is seem unable to accept the fact that evolution is a natural and understandable process, preferring to believe in some hypothetical outside force that mysteriously guides human development. But since 1859, such unsupported assumptions have become increasingly untenable; nature simply does not reveal meaningful patterns of change toward a predestined goal.

In recent decades, many ethical thinkers have rejected the notion that Darwinism undermines human values (and therefore precludes great and tragic art in our time). On the contrary, they argue that, since Darwin, we have been compelled, in art as in life, to mature to finiteness—that is, not simply to realize that there are no credible absolutes in our moral experience (because that mere realization is still a kind of philosophical adolescence) but to accept our finiteness, to rely on our personal resources, and to determine for ourselves our own best values.

After all, Oedipus, Antigone, and Lear are great tragic figures not because they stand in relation to the gods as flies to wanton boys but because of their own decisions and actions—their own impressive humanity. And evolution has not destroyed our wonder at the special value of our own existence; quite the contrary. We are "the most exalted object which we are capable of conceiving," Darwin wrote, adding, "there is grandeur in this view of life."

If we now consider ourselves special because of our highly developed, highly specialized brain, rather than because of some traditional myth—and if we choose scientific fact in preference to superstition—our place in the world nevertheless remains important to us. Modern literature, despite its concern with absurdity, disillusion, black humor, antiheroism, and existential angst, remains human-centered and human-valued. Most recent writers have kept human aspiration as the tacit theme of their work, so that even when they record the bleak tale of our diverse failings, they also limn the ghostly negative of an implicitly positive picture.

A century and a half after *The Origin of Species*, Darwin himself continues to intrigue writers. Contemporary literature, like other current disciplines, has welcomed the larger perspectives and the sharpened perceptions that come from a clear understanding of Darwin's hard-earned reality.

8

We are perpetually moralists, Samuel Johnson said, but only occasionally geometricians; scientists and philosophers of science continue to ponder the origins of our communal ethics. Darwin, as usual, had anticipated them, and his ideas on the subject are still valuable. In *The Descent of Man*, Darwin speculated about our moral as well as our physical origins. He realized, better perhaps than anyone, that at all levels of animal life, the first necessity is self-preservation; but he also recognized that self-preservation sometimes means extending our survival tactics beyond the individual. Even at primitive levels of social organization, human beings have recognized that to live together in communities—as people must, to survive at all—we have to have some basic mutual agreements, some ground rules that we try to abide by. As Darwin observed, these can often be reduced to a kind of "golden rule": treat others as you would like to be treated. The golden rule principle undoubtedly evolved over time as a social necessity, a practical understanding; and its worldwide acknowledgment indicates that it is a general human concept, not the unique property of any one culture or religion.

Contemplating our social ground rules, Darwin surmised that after millions of years of living together in communities, our social behavior might to some extent be inherited. Darwin called that biological legacy "social instinct," the inheritance from our long past not only of the selfpreservation imperative (the so-called animalistic urges that often cause people to be extraordinarily selfish and sometimes unscrupulous) but also our tendencies for "good" social behavior—that is, showing respect for others, fair dealing, honesty, and by a natural extension, kindness and charitableness.

It is commonly assumed that science, as such, has no ethics, that the gap between what "is" and what "ought to be" is broad and unbridgeable. However, scientific knowledge and understanding often do inform our ethical choices. For one thing, scientific knowledge has an important ethical role in identifying false issues that we can and should ignore—such as imagined astrological influences on our moral decisions. Such identification of error is obviously in itself a valuable service.

Beyond that, the growth of scientific knowledge has tended to have socially progressive effects. On the whole, factual knowledge of the physical world has been a far better basis for human understanding, human solidarity, and human sympathy than were folklore or superstition. The old myths of tribal and racial superiority, for instance, have now been thoroughly discredited by the biological understanding that we are one people, one species, in one world.

Moreover, it is the objective scientists who are actually now in the vanguard of ethical thought. In fact, some of the wisest thinkers who have reflected on the human condition over the last century have been biologists—from Darwin's friend T. H. Huxley a century ago to the many present-day zoologists, philosophers, ethologists, and other thinkers who continue to grapple with the complexities of ethical theory and practice, frequently illuminated by the new perspectives of our increasing scientific knowledge.

In his groundbreaking book *The Expression of the Emotions in Man and Animals*, Darwin demonstrated that we share not only our flesh and blood with our fellow creatures but, in varying degrees, our emotional lives as well. With this knowledge, Darwin helped us understand the true wonder and significance of being human—an understanding that has since been broadened and deepened by thousands of hardworking biologists and anthropologists, who continue to examine the real world on our behalf. Thanks to all of them, as well as to Darwin, we now know the true story of our physical and emotional history, which reaches back through the millions of years of human and protohuman development, and even back to our animal ancestry—a history that is implicit in the ancient bones that the paleontologists are constantly bringing forth, and in the genes of every cell of every person on earth.

These basic and essential understandings have been achieved only because biology is a science thoroughly unified by the principle of evolution—and that unification we owe to Charles Darwin. In 1834, when the polymath Samuel Taylor Coleridge died, he believed that the science of zoology was in danger of falling apart because of its huge mass of uncoordinated factual information. That was just four years before Darwin picked up Malthus's *Essay on the Principle of Population* and discovered the clue to natural selection, and thus to the great organizing principle of evolution that would for the first time make a mature and coherent science of biology. Evolution by natural selection continues to serve biologists well in their professional work and in their ongoing explorations of our place in the universe. As Darwin wrote in *The Descent of Man*:

> The moral faculties are generally and justly esteemed as of higher value than the intellectual powers. But . . . the activity of the mind in vividly recalling past impressions is one of the . . . bases of conscience. This affords the strongest argument for educating and stimulating in all possible ways the intellectual faculties of every human being."

The future is by definition obscure and risky, and admonitions about it notoriously fallible. But Darwin's sensible advice—"educating and stimulating in all possible ways the intellectual faculties of every human being"—probably gives us our best chance at what he himself called a "higher destiny in the distant future."

Portrait of Charles Darwin (Bettmann/Corbis)

The Origin of Species (1859)†

Introduction

When on board H.M.S. 'Beagle,' as naturalist, I was much struck with certain facts in the distribution of the inhabitants of South America, and in the geological relations of the present to the past inhabitants of that continent. These facts seemed to me to throw some light on the origin of species—that mystery of mysteries, as it has been called by one of our greatest philosophers. On my return home, it occurred to me, in 1837, that something might perhaps be made out on this question by patiently accumulating and reflecting on all sorts of facts which could possibly have any bearing on it. After five years' work I allowed myself to speculate on the subject, and drew up some short notes; these I enlarged in 1844 into a sketch of the conclusions, which then seemed to me probable: from that period to the present day I have steadily pursued the same object. I hope that I may be excused for entering on these personal details, as I give them to show that I have not been hasty in coming to a decision.

My work is now nearly finished; but as it will take me two or three more years to complete it, and as my health is far from strong, I have been urged to publish this Abstract. I have more especially been induced to do this, as Mr. Wallace, who is now studying the natural history of the Malay archipelago, has arrived at almost exactly the same general conclusions that I have on the origin of species. Last year he sent to me a memoir on this subject, with a request that I would forward it to Sir Charles Lyell, who sent it to the Linnean Society, and it is published in the third volume of the Journal of that Society. Sir C. Lyell and Dr. Hooker, who both knew of my work—the latter having read my sketch of 1844—honoured me by thinking it advisable to publish, with Mr. Wallace's excellent memoir, some brief extracts from my manuscripts.

This Abstract, which I now publish, must necessarily be imperfect. I cannot here give references and authorities for my several statements; and I must trust to the reader reposing some confidence in my accuracy. No doubt errors will have crept in, though I hope I have always been cautious in trusting to good authorities alone. I can here give only the general conclusions at which I have arrived, with a few facts in illustration, but which, I hope, in most cases

† The text is excerpted from the first edition of *The Origin of Species* (London, 1859). Darwin's title for the first edition of the book was *On the Origin of Species*, but he dropped the *On* in the sixth (and last) edition of his lifetime, and the shorter title has been commonly used ever since.

will suffice. No one can feel more sensible than I do of the necessity of hereafter publishing in detail all the facts, with references, on which my conclusions have been grounded; and I hope in a future work to do this. For I am well aware that scarcely a single point is discussed in this volume on which facts cannot be adduced, often apparently leading to conclusions directly opposite to those at which I have arrived. A fair result can be obtained only by fully stating and balancing the facts and arguments on both sides of each question; and this cannot possibly be here done.

I much regret that want of space prevents my having the satisfaction of acknowledging the generous assistance which I have received from very many naturalists, some of them personally unknown to me. I cannot, however, let this opportunity pass without expressing my deep obligations to Dr. Hooker, who for the last fifteen years has aided me in every possible way by his large stores of knowledge and his excellent judgment.

In considering the Origin of Species, it is quite conceivable that a naturalist, reflecting on the mutual affinities of organic beings, on their embryological relations, their geographical distribution, geological succession, and other such facts, might come to the conclusion that each species had not been independently created, but had descended, like varieties, from other species. Nevertheless, such a conclusion, even if well founded, would be unsatisfactory, until it could be shown how the innumerable species inhabiting this world have been modified, so as to acquire that perfection of structure and coadaptation which most justly excites our admiration. Naturalists continually refer to external conditions, such as climate, food, &c., as the only possible cause of variation. In one very limited sense, as we shall hereafter see, this may be true; but it is preposterous to attribute to mere external conditions, the structure, for instance, of the woodpecker, with its feet, tail, beak, and tongue, so admirably adapted to catch insects under the bark of trees. In the case of the misseltoe, which draws its nourishment from certain trees, which has seeds that must be transported by certain birds, and which has flowers with separate sexes absolutely requiring the agency of certain insects to bring pollen from one flower to the other, it is equally preposterous to account for the structure of this parasite, with its relations to several distinct organic beings, by the effects of external conditions, or of habit, or of the volition of the plant itself.

The author of the 'Vestiges of Creation' would, I presume, say that, after a certain unknown number of generations, some bird had given birth to a woodpecker, and some plant to the misseltoe, and that these had been produced perfect as we now see them; but this assumption seems to me to be no explanation, for it leaves the case of the coadaptations of organic beings to each other and to their physical conditions of life, untouched and unexplained.

It is, therefore, of the highest importance to gain a clear insight into the means of modification and coadaptation. At the commencement of my observations it seemed to me probable that a careful study of domesticated animals and of cultivated plants would offer the best chance of making out this obscure problem. Nor have I been disappointed; in this and in all other perplexing cases I have invariably found that our knowledge, imperfect though it be, of variation under domestication, afforded the best and safest clue. I may venture to express my conviction of the high value of such studies, although they have been very commonly neglected by naturalists.

From these considerations, I shall devote the first chapter of this Abstract to Variation under Domestication. We shall thus see that a large amount of hereditary modification is at least possible; and, what is equally or more important, we shall see how great is the power of man in accumulating by his Selection successive slight variations. I will then pass on to the variability of species in a state of nature; but I shall, unfortunately, be compelled to treat this subject far too briefly, as it can be treated properly only by giving long catalogues of facts. We shall, however, be enabled to discuss what circumstances are most favourable to variation. In the next chapter the Struggle for Existence amongst all organic beings throughout the world, which inevitably follows from their high geometrical powers of increase, will be treated of. This is the doctrine of Malthus, applied to the whole animal and vegetable kingdoms. As many more individuals of each species are born than can possibly survive; and as, consequently, there is a frequently recurring struggle for existence, it follows that any being, if it vary however slightly in any manner profitable to itself, under the complex and sometimes varying conditions of life, will have a better chance of surviving, and thus be *naturally selected*. From the strong principle of inheritance, any selected variety will tend to propagate its new and modified form.

This fundamental subject of Natural Selection will be treated at some length in the fourth chapter; and we shall then see how Natural Selection almost inevitably causes much Extinction of the less improved forms of life, and induces what I have called Divergence of Character. In the next chapter I shall discuss the complex and little known laws of variation and of correlation of growth. In the four succeeding chapters, the most apparent and gravest difficulties on the theory will be given: namely, first, the difficulties of transitions, or in understanding how a simple being or a simple organ can be changed and perfected into a highly developed being or elaborately constructed organ; secondly, the subject of Instinct, or the mental powers of animals; thirdly, Hybridism, or the infertility of species and the fertility of varieties when intercrossed; and fourthly, the imperfection of the Geological Record. In the next chapter I shall consider the geological succession of organic beings throughout time; in the eleventh and twelfth, their geographical distribution throughout space; in the thirteenth, their classification or mutual affinities, both when mature and in an embryonic condition. In the last chapter I shall give a brief recapitulation of the whole work, and a few concluding remarks.

No one ought to feel surprise at much remaining as yet unexplained in regard to the origin of species and varieties, if he makes due allowance for our profound ignorance in regard to the mutual relations of all the beings which live around us. Who can explain why one species ranges widely and is very numerous, and why another allied species has a narrow range and is rare? Yet these relations are of the highest importance, for they determine the present welfare, and, as I believe, the future success and modification of every inhabitant of this world. Still less do we know of the mutual relations of the innumerable inhabitants of the world during the many past geological epochs in its history. Although much remains obscure, and will long remain obscure, I can entertain no doubt, after the most deliberate study and dispassionate judgment of which I am capable, that the view which most naturalists entertain, and which I formerly entertained—namely, that each species has been independently created—is erroneous. I am fully convinced that species are not immutable; but that those belonging to what are called the same genera are lineal descendants of some other and generally extinct species, in the same manner as the acknowledged varieties of any one species are the descendants of that species. Furthermore, I am convinced that Natural Selection has been the main but not exclusive means of modification.

Chapter I. Variation under Domestication

Causes of Variability—Effects of Habit—Correlation of Growth—Inheritance—Character of Domestic Varieties—Difficulty of distinguishing between Varieties and Species—Origin of Domestic Varieties from one or more Species—Domestic Pigeons, their Differences and Origin—Principle of Selection anciently followed, its Effects—Methodical and Unconscious Selection—Unknown Origin of our Domestic Productions—Circumstances favourable to Man's power of Selection.

When we look to the individuals of the same variety or sub-variety of our older cultivated plants and animals, one of the first points which strikes us, is, that they generally differ much more from each other, than do the individuals of any one species or variety in a state of nature. When we reflect on the vast diversity of the plants and animals which have been cultivated, and which have varied during all ages under the most different climates and treatment, I think we are driven to conclude that this greater variability is simply due to our domestic productions having been raised under conditions of life not so uniform as, and somewhat different from, those to which the parent-species have been exposed under nature. There is, also, I think, some probability in the view propounded by Andrew Knight, that this variability may be partly connected with excess of food. It seems pretty clear that organic beings must be exposed during several generations to the new conditions of life to cause any appreciable amount of variation; and that when the organisation has once begun to vary, it generally continues to vary for many generations. No case is on record of a variable being ceasing to be variable under cultivation. Our oldest cultivated plants, such as wheat, still often yield new varieties: our oldest domesticated animals are still capable of rapid improvement or modification.

* * *

The result of the various, quite unknown, or dimly seen laws of variation is infinitely complex and diversified. It is well worth while carefully to study the several treatises published on some of our old cultivated plants, as on the hyacinth, potato, even the dahlia, &c.; and it is really surprising to note the endless points in structure and constitution in which the varieties and subvarieties differ slightly from each other. The whole organisation seems to have become plastic, and tends to depart in some small degree from that of the parental type.

Any variation which is not inherited is unimportant for us. But the number and diversity of inheritable deviations of structure, both those of slight and those of considerable physiological importance, is endless. Dr. Prosper Lucas's treatise, in two large volumes, is the fullest and the best on this subject. No breeder doubts how strong is the tendency to inheritance: like produces like is his fundamental belief: doubts have been thrown on this principle by theoretical writers alone. When a deviation appears not unfrequently, and we see it in the father and child, we cannot tell whether it may not be due to the same original cause acting on both; but when amongst individuals, apparently exposed to the same conditions, any very rare deviation, due to some extraordinary combination of circumstances, appears in the parent—say, once amongst several million individuals—and it reappears in the child, the mere doctrine of chances almost compels us to attribute its reappearance to inheritance. Every one must have heard of cases of albinism, prickly skin, hairy bodies, &c., appearing in several members of the same family. If strange and rare deviations of structure are truly inherited, less strange and commoner deviations may be freely admitted to be inheritable. Perhaps the correct way of viewing the whole subject, would be, to look at the inheritance of every character whatever as the rule, and non-inheritance as the anomaly.

The laws governing inheritance are quite unknown; no one can say why the same peculiarity in different individuals of the same species, and in individuals of different species, is sometimes inherited and sometimes not so; why the child often reverts in certain characters to its grandfather or grandmother or other much more remote ancestor; why a peculiarity is often transmitted from one sex to both sexes, or to one sex alone, more commonly but not exclusively to the like sex. It is a fact of some little importance to us, that peculiarities appearing in the males of our domestic breeds are often transmitted either exclusively, or in a much greater degree, to males alone. A much more important rule, which I think may be trusted, is that, at whatever period of life a peculiarity first appears, it tends to appear in the offspring at a corresponding age, though sometimes earlier. In many cases this could not be otherwise: thus the inherited peculiarities in the horns of cattle could appear only in the offspring when nearly mature; peculiarities in the silkworm are known to appear at the corresponding caterpillar or cocoon stage. But hereditary diseases and some other facts make me believe that the rule has a wider extension, and that when there is no apparent reason why a peculiarity should appear at any particular age, yet that it does tend to appear in the offspring at the same period at which it first appeared in the parent. I believe this rule to be of the highest importance in explaining the laws of embryology. These remarks are of course confined to the first *appearance* of the peculiarity, and not to its primary cause, which may have acted on the ovules or male element; in nearly the same manner as in the crossed offspring from a short-horned cow by a long-horned bull, the greater length of horn, though appearing late in life, is clearly due to the male element.

* * *

When we look to the hereditary varieties or races of our domestic animals and plants, and compare them with species closely allied together, we generally perceive in each domestic race, as already remarked, less uniformity of character than in true species. Domestic races of the same species, also, often have a somewhat monstrous character; by which I mean, that, although differing from each other, and from the other species of the same genus, in several trifling respects, they often differ in an extreme degree in some one part, both when compared one with another, and more especially when compared with all the species in nature to which they are nearest allied. With these exceptions (and with that of the perfect fertility of varieties when crossed,—a subject hereafter to be discussed), domestic races of the same species differ from each other in the same manner as, only in most cases in a lesser degree than, do closely-allied species of the same genus in a state of nature. I think this must be admitted, when we find that there are hardly any domestic races, either amongst animals or plants, which have not been ranked by some competent judges as mere varieties, and by other competent judges as the descendants of aboriginally distinct species. If any marked distinction existed between domestic races and species, this source of doubt could not so perpetually recur. It has often been stated that domestic races do not differ from each other in characters of generic value. I think it could be shown that this statement is hardly correct; but naturalists differ most widely in determining what characters are of generic value; all such valuations being at present empirical. Moreover, on the view of the origin of genera which I shall presently give, we have no right to expect often to meet with generic differences in our domesticated productions.

When we attempt to estimate the amount of structural difference between the domestic races of the same species, we are soon involved in doubt, from not knowing whether they have descended from one or several parent-species. This point, if it could be cleared up, would be interesting; if, for instance, it could be shown that the greyhound, bloodhound, terrier, spaniel, and bull-dog, which we all know propagate their kind so truly, were the offspring of any single species, then such facts would have great weight in making us doubt about the immutability of the many very closely allied and natural species—for instance, of the many foxes—inhabiting different quarters of the world. I do not believe, as we shall presently see, that all our dogs have descended from any one wild species; but, in the case of some other domestic races, there is presumptive, or even strong, evidence in favour of this view.

* * *

ON THE BREEDS OF THE DOMESTIC PIGEON

Believing that it is always best to study some special group, I have, after deliberation, taken up domestic pigeons. I have kept every breed which I could purchase or obtain, and have been most kindly favoured with skins from several quarters of the world, more especially by the Hon. W. Elliot from India, and by the Hon. C. Murray from Persia. Many treatises in different languages have been published on pigeons, and some of them are very important, as being of considerable antiquity. I have associated with several eminent fanciers, and have been permitted to join two of the London Pigeon Clubs. The diversity of the breeds is something astonishing. Compare the English carrier and the short-faced tumbler, and see the wonderful difference in their beaks, entailing corresponding differences in their skulls. The carrier, more especially the male bird, is also remarkable from the wonderful development of the carunculated skin about the head, and this is accompanied by greatly elongated eyelids, very large external orifices to the nostrils, and a wide gape of mouth. The short-faced tumbler has a beak in outline almost like that of a finch; and the common tumbler has the singular and strictly inherited habit of flying at a great height in a compact flock, and tumbling in the air head over heels. The runt is a bird of great size, with long, massive beak and large feet; some of the sub-breeds of runts have very long necks, others very long wings and tails, others singularly short tails. The barb is allied to the carrier, but, instead of a very long beak, has a very short and very broad one. The pouter has a much elongated body, wings, and legs; and its enormously developed crop, which it glories in inflating, may well excite astonishment and even laughter. The turbit has a very short and conical beak, with a line of reversed feathers down the breast; and it has the habit of continually expanding slightly the upper part of the œsophagus. The Jacobin has the feathers so much reversed along the back of the neck that they form a hood, and it has, proportionally to its size, much elongated wing and tail feathers. The trumpeter and laugher, as their names express, utter a very different coo from the other breeds. The fantail has thirty or even forty tail-feathers, instead of twelve or fourteen, the normal number in all members of the great pigeon family; and these feathers are kept expanded, and are carried so erect that in good birds the head and tail touch; the oil-gland is quite aborted. Several other less distinct breeds might have been specified.

In the skeletons of the several breeds, the development of the bones of the face in length and breadth and curvature differs enormously. The shape, as well as the breadth and length of the ramus of the lower jaw, varies in a highly remarkable manner. The number of the caudal and sacral vertebræ vary; as does the number of the ribs, together with their relative breadth and the presence of processes. The size and shape of the apertures in the sternum are highly variable; so is the degree of divergence and relative size of the two arms of the furcula. The proportional width of the gape of mouth, the proportional length of the eyelids, of the orifice of the nostrils, of the tongue (not always in strict correlation with the length of beak), the size of the crop and of the upper part of the œsophagus; the development and abortion of the oil-gland; the number of the primary wing and caudal feathers; the relative length of wing and tail to each other and to the body; the relative length of leg and of the feet; the number of scutellæ on the toes, the development of skin between the toes, are all points of structure which are variable. The period at which the perfect plumage is acquired varies, as does the state of the down with which the nestling birds are clothed when hatched. The shape and size of the eggs vary. The manner of flight differs remarkably; as does in some breeds the voice and disposition. Lastly, in certain breeds, the males and females have come to differ to a slight degree from each other.

Altogether at least a score of pigeons might be chosen, which if shown to an ornithologist, and he were told that they were wild birds, would certainly, I think, be ranked by him as well-defined species. Moreover, I do not believe that any ornithologist would place the English carrier, the short-faced tumbler, the runt, the barb, pouter, and fantail in the same genus; more especially as in each of these breeds several truly-inherited sub-breeds, or species as he might have called them, could be shown him.

Great as the differences are between the breeds of pigeons, I am fully convinced that the common opinion of naturalists is correct, namely, that all have descended from the rock-pigeon (Columba livia), including under this term several geographical races or sub-species, which differ from each other in the most trifling respects.

* * *

From these several reasons, namely, the improbability of man having formerly got seven or eight supposed species of pigeons to breed freely under domestication; these supposed species being quite unknown in a wild state, and their becoming nowhere feral; these species having very abnormal characters in certain respects, as compared with all other Columbidæ, though so like in most other respects to the rock-pigeon; the blue colour and various marks occasionally appearing in all the breeds, both when kept pure and when crossed; the mongrel offspring being perfectly fertile;—from these several reasons, taken together, I can feel no doubt that all our domestic breeds have descended from the Columba livia with its geographical sub-species.

In favour of this view, I may add, firstly, that C. livia, or the rock-pigeon, has been found capable of domestication in Europe and in

India; and that it agrees in habits and in a great number of points of structure with all the domestic breeds. Secondly, although an English carrier or short-faced tumbler differs immensely in certain characters from the rock-pigeon, yet by comparing the several sub-breeds of these breeds, more especially those brought from distant countries, we can make an almost perfect series between the extremes of structure. Thirdly, those characters which are mainly distinctive of each breed, for instance the wattle and length of beak of the carrier, the shortness of that of the tumbler, and the number of tail-feathers in the fantail, are in each breed eminently variable; and the explanation of this fact will be obvious when we come to treat of selection. Fourthly, pigeons have been watched, and tended with the utmost care, and loved by many people. They have been domesticated for thousands of years in several quarters of the world; the earliest known record of pigeons is in the fifth Ægyptian dynasty, about 3000 B.C., as was pointed out to me by Professor Lepsius; but Mr. Birch informs me that pigeons are given in a bill of fare in the previous dynasty. In the time of the Romans, as we hear from Pliny, immense prices were given for pigeons; "nay, they are come to this pass, that they can reckon up their pedigree and race." Pigeons were much valued by Akber Khan in India, about the year 1600; never less than 20,000 pigeons were taken with the court. "The monarchs of Iran and Turan sent him some very rare birds;" and, continues the courtly historian, "His Majesty by crossing the breeds, which method was never practised before, has improved them astonishingly." About this same period the Dutch were as eager about pigeons as were the old Romans. The paramount importance of these considerations in explaining the immense amount of variation which pigeons have undergone, will be obvious when we treat of Selection. We shall then, also, see how it is that the breeds so often have a somewhat monstrous character. It is also a most favourable circumstance for the production of distinct breeds, that male and female pigeons can be easily mated for life; and thus different breeds can be kept together in the same aviary.

I have discussed the probable origin of domestic pigeons at some, yet quite insufficient, length; because when I first kept pigeons and watched the several kinds, knowing well how true they bred, I felt fully as much difficulty in believing that they could ever have descended from a common parent, as any naturalist could in coming to a similar conclusion in regard to the many species of finches, or other large groups of birds, in nature. One circumstance has struck me much; namely, that all the breeders of the various domestic animals and the cultivators of plants, with whom I have ever conversed, or whose treatises I have read, are firmly convinced that the several breeds to which each has attended, are descended from so many aboriginally distinct species. Ask, as I have asked, a celebrated raiser of Hereford cattle, whether his cattle might not have descended from long-horns, and he will laugh you to scorn. I have never met a pigeon, or poultry, or duck, or rabbit fancier, who was not fully convinced that each main breed was descended from a distinct species. Van Mons, in his treatise on pears and apples, shows how utterly he disbelieves that the several sorts, for instance a Ribston-pippin or Codlin-apple, could ever have proceeded from the seeds of the same tree. Innumerable other examples could be given. The explanation, I think, is simple: from long-continued study they are strongly impressed with the differences between the several races; and though they well know that each race varies slightly, for they win their prizes by selecting such slight differences, yet they ignore all general arguments, and refuse to sum up in their minds slight differences accumulated during many successive generations. May not those naturalists who, knowing far less of the laws of inheritance than does the breeder, and knowing no more than he does of the intermediate links in the long lines of descent, yet admit that many of our domestic races have descended from the same parents—may they not learn a lesson of caution, when they deride the idea of species in a state of nature being lineal descendants of other species?

SELECTION

Let us now briefly consider the steps by which domestic races have been produced, either from one or from several allied species. Some little effect may, perhaps, be attributed to the direct action of the external conditions of life, and some little to habit; but he would be a bold man who would account by such agencies for the differences of a dray and race horse, a greyhound and bloodhound, a carrier and tumbler pigeon. One of the most remarkable features in our domesticated races is that we see in them adaptation, not indeed to the animal's or plant's own good, but to man's use or fancy. Some variations useful to him have probably arisen suddenly, or by one step; many botanists, for instance, believe that the fuller's teazle, with its hooks, which cannot be rivalled by any mechanical contrivance, is only a variety of the wild Dipsacus; and this amount of change may have suddenly arisen in a seedling. So it has probably been with the turnspit dog; and this is known to have been the case with the ancon sheep. But when we compare the dray-horse and race-horse, the dromedary and camel, the various breeds of sheep fitted either for cultivated land or mountain pasture, with the wool of one breed good for one purpose, and that of another breed for another purpose; when we compare the many breeds of dogs, each good for man in very different ways; when we compare the game-cock, so pertinacious in battle, with other breeds so little quarrelsome, with "everlasting layers" which never desire to sit, and with the bantam so small and elegant; when we compare the host of agricultural, culinary, orchard, and flower-garden races of plants, most useful to man at different seasons and for different purposes, or so beautiful in his eyes, we must, I think, look further than to mere variability. We cannot suppose that all the breeds were suddenly produced as perfect and as useful as we now see them; indeed, in several cases, we know that this has not been their history. The key is man's power of accumulative selection: nature gives successive variations; man adds them up in certain directions useful to him. In this sense he may be said to make for himself useful breeds.

* * *

At the present time, eminent breeders try by methodical selection, with a distinct object in view, to make a new strain or sub-breed, superior to anything existing in the country. But, for our purpose, a kind of Selection, which may be called Unconscious, and which results from every one trying to possess and breed from the best individual animals, is more important. Thus, a man who intends keeping pointers naturally tries to get as good dogs as he can, and afterwards breeds from his own best dogs, but he has no wish or expectation of permanently altering the breed. Nevertheless I cannot doubt that this process, continued during centuries, would improve and modify any breed, in the same way as Bakewell, Collins, &c., by this very same process, only carried on more methodically, did greatly modify, even during their own lifetimes, the forms and qualities of their cattle. Slow and insensible changes of this kind could never be recognised unless actual measurements or careful drawings of the breeds in question had been made long ago, which might serve for comparison. In some cases, however, unchanged or but little changed individuals of the same breed may be found in less civilised districts, where the breed has been less improved. There is reason to believe that King Charles's spaniel has been unconsciously modified to a large extent since the time of that monarch. Some highly competent authorities are convinced that the setter is directly derived from the spaniel, and has probably been slowly altered from it. It is known that the English pointer has been greatly changed within the last century, and in this case the change has, it is believed, been chiefly effected by crosses with the foxhound; but what concerns us is, that the change has been effected unconsciously and gradually, and yet so effectually, that, though the old Spanish pointer certainly came from Spain, Mr. Borrow has not seen, as I am informed by him, any native dog in Spain like our pointer.

By a similar process of selection, and by careful training, the whole body of English racehorses have come to surpass in fleetness and size the parent Arab stock, so that the latter, by the regulations for the Goodwood Races, are favoured in the weights they carry. Lord Spencer and others have shown how the cattle of England have increased in weight and in early maturity, compared with the stock formerly kept in this country. By comparing the accounts given in old pigeon treatises of carriers and tumblers with these breeds as now existing in Britain, India, and Persia, we can, I think, clearly trace the stages through which they have insensibly passed, and come to differ so greatly from the rock-pigeon.

* * *

To sum up on the origin of our Domestic Races of animals and plants. I believe that the conditions of life, from their action on the reproductive system, are so far of the highest importance as causing variability. I do not believe that variability is an inherent and necessary contingency, under all circumstances, with all organic beings, as some authors have thought. The effects of variability are modified by various degrees of inheritance and of reversion. Variability is governed by many unknown laws, more especially by that of correlation of growth. Something may be attributed to the direct action of the conditions of life. Something must be attributed to use and disuse. The final result is thus rendered infinitely complex. In some cases, I do not doubt that the intercrossing of species, aboriginally distinct, has played an important part in the origin of our domestic productions. When in any country several domestic breeds have once been established, their occasional intercrossing, with the aid of selection, has, no doubt, largely aided in the formation of new sub-breeds; but the importance of the crossing of varieties has, I believe, been greatly exaggerated, both in regard to animals and to those plants which are propagated by seed. In plants which are temporarily propagated by cuttings, buds, &c., the importance of the crossing both of distinct species and of varieties is immense; for the cultivator here quite disregards the extreme variability both of hybrids and mongrels, and the frequent sterility of hybrids; but the cases of plants not propagated by seed are of little importance to us, for their endurance is only temporary. Over all these causes of Change I am convinced that the accumulative action of Selection, whether applied methodically and more quickly, or unconsciously and more slowly, but more efficiently, is by far the predominant Power.

Chapter II. Variation under Nature

Variability—Individual differences—Doubtful species—Wide ranging, much diffused, and common species vary most—Species of the larger genera in any country vary more than the species of the smaller genera—Many of the species of the larger genera resemble varieties in being very closely, but unequally, related to each other, and in having restricted ranges.

Before applying the principles arrived at in the last chapter to organic beings in a state of nature, we must briefly discuss whether these latter are subject to any variation. To treat this subject at all properly, a long catalogue of dry facts should be given; but these I shall reserve for my future work. Nor shall I here discuss the various definitions which have been given of the term species. No one definition has as yet satisfied all naturalists; yet every naturalist knows vaguely what he means when he speaks of a species. Generally the term includes the unknown element of a distinct act of creation. The term "variety" is almost equally difficult to define; but here community of descent is almost universally implied, though it can rarely be proved.

* * *

Alph. De Candolle and others have shown that plants which have very wide ranges generally present varieties; and this might have been expected, as they become exposed to diverse physical conditions, and as they come into competition (which, as we shall hereafter see, is a far more important circumstance) with different sets of organic beings. But my tables further show that, in any limited country, the species which are most common, that is abound most in individuals, and the species which are most widely diffused within their own country (and this is a different consideration from wide range, and to a certain extent from commonness), often give rise to varieties sufficiently well-marked to have been recorded in botanical works. Hence it is the most flourishing, or, as they may be called, the dominant species,—those which range widely over the world, are the most diffused in their own country, and are the most numerous in individuals,—which oftenest produce well-marked varieties, or, as I consider them, incipient species. And this, perhaps, might have been anticipated; for, as varieties, in order to become in any degree permanent, necessarily have to struggle with the other inhabitants of the country, the species which are already dominant will be the most likely to yield offspring which, though in some slight degree modified, will still inherit those advantages that enabled their parents to become dominant over their compatriots.

* * *

Finally, then, varieties have the same general characters as species, for they cannot be distinguished from species,—except, firstly, by the discovery of intermediate linking forms, and the occurrence of such links cannot affect the actual characters of the forms which they connect; and except, secondly, by a certain amount of difference, for two forms, if differing very little, are generally ranked as varieties, notwithstanding that intermediate linking forms have not been discovered; but the amount of difference considered necessary to give to two forms the rank of species is quite indefinite. In genera having more than the average number of species in any country, the species of these genera have more than the average number of varieties. In large genera the species are apt to be closely, but unequally, allied together, forming little clusters round certain species. Species very closely allied to other species apparently have restricted ranges. In all these several respects the species of large genera present a strong analogy with varieties. And we can clearly understand these analogies, if species have once existed as varieties, and have thus originated: whereas, these analogies are utterly inexplicable if each species has been independently created.

We have, also, seen that it is the most flourishing and dominant species of the larger genera which on an average vary most; and varieties, as we shall hereafter see, tend to become converted into new and distinct species. The larger genera thus tend to become larger; and throughout nature the forms of life which are now dominant tend to become still more dominant by leaving many modified and dominant descendants. But by steps hereafter to be explained, the larger genera also tend to break up into smaller genera. And thus, the forms of life throughout the universe become divided into groups subordinate to groups.

Chapter III. Struggle for Existence

Bears on natural selection—The term used in a wide sense—Geometrical powers of increase—Rapid increase of naturalised animals and plants—Nature of the checks to increase—Competition universal—Effects of climate—Protection from the number of individuals—Complex relations of all animals and plants throughout nature—Struggle for life most severe between individuals and varieties of the same species; often severe between species of the same genus—The relation of organism to organism the most important of all relations.

Before entering on the subject of this chapter, I must make a few preliminary remarks, to show how the struggle for existence bears on Natural Selection. It has been seen in the last chapter that amongst organic beings in a state of nature there is some individual variability; indeed I am not aware that this has ever been disputed. It is immaterial for us whether a multitude of doubtful forms be called species or sub-species or varieties; what rank, for instance, the two or three hundred doubtful forms of British plants are entitled to hold, if the existence of any well-marked varieties be admitted. But the mere existence of individual variability and of some few well-marked varieties, though necessary as the foundation for the work, helps us but little in understanding how species arise in nature. How have all those exquisite adaptations of one part of the organisation to another part, and to the conditions of life, and of one distinct organic being to another being, been perfected? We see these beautiful co-adaptations most plainly in the woodpecker and missletoe; and only a little less plainly in the humblest parasite which clings to the hairs of a quadruped or feathers of a bird; in the structure of the beetle which dives through the water; in the plumed seed which is wafted by the gentlest breeze; in short, we see beautiful adaptations everywhere and in every part of the organic world.

Again, it may be asked, how is it that varieties, which I have called incipient species, become ultimately converted into good and distinct species, which in most cases obviously differ from each other far more than do the varieties of the same species? How do those groups of species, which constitute what are called distinct genera, and which differ from each other more than do the species of the same genus, arise? All these results, as we shall more fully see in the next chapter, follow inevitably from the struggle for life. Owing to this struggle for life, any variation, however slight and from whatever cause proceeding, if it be in any degree profitable to an individual of any species, in its infinitely complex relations to other organic beings and to external nature, will tend to the preservation of that individual, and will generally be inherited by its offspring. The offspring, also, will thus have a better chance of surviving, for, of the many individuals of any species which are periodically born, but a small number can survive. I have called this principle, by which each slight variation, if useful, is preserved, by the term of Natural Selection, in order to mark its relation to man's power of selection. We have seen that man by selection can certainly produce great results, and can adapt organic beings to his own uses, through the accumulation of slight but useful variations, given to him by the hand of Nature. But Natural Selection, as we shall hereafter see, is a power incessantly ready for action, and is as immeasurably superior to man's feeble efforts, as the works of Nature are to those of Art.

We will now discuss in a little more detail the struggle for existence. In my future work this subject shall be treated, as it well deserves, at much greater length. The elder De Candolle and Lyell have largely and philosophically shown that all organic beings are exposed to severe competition. In regard to plants, no one has treated this subject with more spirit and ability than W. Herbert, Dean of Manchester, evidently the result of his great horticultural knowledge. Nothing is easier than to admit in words the truth of the universal struggle for life, or more difficult—at least I have found it so—than constantly to bear this conclusion in mind. Yet unless it be thoroughly engrained in the mind, I am convinced that the whole economy of nature, with every fact on distribution, rarity, abundance, extinction, and variation, will be dimly seen or quite misunderstood. We behold the face of nature bright with gladness, we often see superabundance of food; we do not see, or we forget, that the birds which are idly singing round us mostly live on insects or seeds, and are thus constantly destroying life; or we forget how largely these songsters, or their eggs, or their nestlings, are destroyed by birds and beasts of prey; we do not always bear in mind, that though food may be now superabundant, it is not so at all seasons of each recurring year.

I should premise that I use the term Struggle for Existence in a large and metaphorical sense, including dependence of one being on another, and including (which is more important) not only the life of the individual, but success in leaving progeny. Two canine animals in a time of dearth, may be truly said to struggle with each other which shall get food and live. But a plant on the edge of a desert is said to struggle for life against the drought, though more properly it should be said to be dependent on the moisture. A plant which annually produces a thousand seeds, of which on an average only one comes to maturity, may be more truly said to struggle with the plants of the same and other kinds which already clothe the ground. The missletoe is dependent on the apple and a few other trees, but can only in a far-fetched sense be said to struggle with these trees, for if too many of these parasites grow on the same tree, it will languish and die. But several seedling missletoes, growing close together on the same branch, may more truly be said to struggle with each other. As the missletoe is disseminated by birds, its existence depends on birds; and it may metaphorically be said to struggle with other fruit-bearing plants, in order to tempt birds to devour and thus disseminate its seeds rather than those of other plants. In these several senses, which pass into each other, I use for convenience sake the general term of struggle for existence.

A struggle for existence inevitably follows from the high rate at which all organic beings tend to increase. Every being, which during its natural lifetime produces several eggs or seeds, must suffer destruction during some period of its life, and during some season or occasional year, otherwise, on the principle of geometrical increase, its numbers would quickly become so inordinately great that no country could support the product. Hence, as more individuals are produced than can possibly survive, there must in every case be a struggle for existence, either one individual with another of the same species, or with the individuals of distinct species, or with the physical conditions of life. It is the doctrine of Malthus applied with manifold force to the whole animal and vegetable kingdoms; for in this case there can be no artificial increase of food, and no prudential restraint from marriage. Although some species may be now increasing, more or less rapidly, in numbers, all cannot do so, for the world would not hold them.

There is no exception to the rule that every organic being naturally increases at so high a rate, that if not destroyed, the earth would soon be covered by the progeny of a single pair. Even slow-breeding man has doubled in twenty-five years, and at this rate, in a few thousand years, there would literally not be standing room for his progeny. Linnæus has calculated that if an annual plant produced only two seeds—and there is no plant so unproductive as this—and their seedlings next year produced two, and so on, then in twenty years there would be a million plants. The elephant is reckoned to be the slowest breeder of all known animals, and I have taken some pains to estimate its probable minimum rate of natural increase: it will be under the mark to assume that it breeds when thirty years old, and goes on breeding till ninety years old, bringing forth three pair of young in this interval; if this be so, at the end of the fifth century there would be alive fifteen million elephants, descended from the first pair.

But we have better evidence on this subject than mere theoretical calculations, namely, the numerous recorded cases of the astonishingly rapid increase of various animals in a state of nature, when circumstances have been favourable to them during two or three following seasons. Still more striking is the evidence from our domestic animals of many kinds which have run wild in several parts of the world: if the statements of the rate of increase of slow-breeding cattle and horses in South-America, and latterly in Australia, had not been well authenticated, they would have been quite incredible. So it is with plants: cases could be given of introduced plants which have become common throughout whole islands in a period of less than ten years. Several of the plants now most numerous over the wide plains of La Plata, clothing square leagues of surface almost to the exclusion of all other plants, have been introduced from Europe; and there are plants which now range in India, as I hear from Dr. Falconer, from Cape Comorin to the Himalaya, which have been imported from America since its discovery. In such cases, and endless instances could be given, no one supposes that the fertility of these animals or plants has been suddenly and temporarily increased in any sensible degree. The obvious explanation is that the conditions of life have been very favourable, and that there has consequently been less destruction of the old and young, and that nearly all the young have been enabled to breed. In such cases the geometrical ratio of increase, the result of which never fails to be surprising, simply explains the extraordinarily rapid increase and wide diffusion of naturalised productions in their new homes.

In a state of nature almost every plant produces seed, and amongst animals there are very few which do not annually pair. Hence we may confidently assert, that all plants and animals are tending to increase at a geometrical ratio, that all would most rapidly stock every station in which they could any how exist, and that the geometrical tendency to increase must be checked by destruction at some period of life. Our familiarity with the larger domestic animals tends, I think, to mislead us: we see no great destruction falling on them, and we forget that thousands are annually slaughtered for food, and that in a state of nature an equal number would have somehow to be disposed of.

The only difference between organisms which annually produce eggs or seeds by the thousand, and those which produce extremely few, is, that the slow-breeders would require a few more years to people, under favourable conditions, a whole district, let it be ever so large. The condor lays a couple of eggs and the ostrich a score, and yet in the same country the condor may be the more numerous of the two: the Fulmar petrel lays but one egg, yet it is believed to be the most numerous bird in the world. One fly deposits hundreds of eggs, and another, like the hippobosca, a single one; but this difference does not determine how many individuals of the two species can be supported in a district. A large number of eggs is of some importance to those species, which depend on a rapidly fluctuating amount of food, for it allows them rapidly to increase in number. But the real importance of a large number of eggs or seeds is to make up for much destruction at some period of life; and this period in the great majority of cases is an early one. If an animal can in any way protect its own eggs or young, a small number may be produced, and yet the average stock be fully kept up; but if many eggs or young are destroyed, many must be produced, or the species will become extinct. It would suffice to keep up the full number of a tree, which lived on an average for a thousand years, if a single seed were produced once in a thousand years, supposing that this seed were never destroyed, and could be ensured to germinate in a fitting place. So that in all cases, the average number of any animal or plant depends only indirectly on the number of its eggs or seeds.

In looking at Nature, it is most necessary to keep the foregoing considerations always in mind—never to forget that every single organic being around us may be said to be striving to the utmost to increase in numbers; that each lives by a struggle at some period of its life; that heavy destruction inevitably falls either on the young or old, during each generation or at recurrent intervals. Lighten any check, mitigate the destruction ever so little, and the number of the species will almost instantaneously increase to any amount. The face of Nature may be compared to a yielding surface, with ten thousand sharp wedges packed close together and driven inwards by incessant blows, sometimes one wedge being struck, and then another with greater force.

* * *

Chapter IV. Natural Selection

Natural Selection—its power compared with man's selection—its power on characters of trifling importance—its power at all ages and on both sexes—Sexual Selection—On the generality of intercrosses between individuals of the same species—Circumstances favourable and unfavourable to Natural Selection, namely, intercrossing, isolation, number of individuals—Slow action—Extinction caused by Natural Selection—Divergence of Character, related to the diversity of inhabitants of any small area, and to naturalisation—Action of Natural Selection, through Divergence of Character and Extinction, on the descendants from a common parent—Explains the Grouping of all organic beings.

How will the struggle for existence, discussed too briefly in the last chapter, act in regard to variation? Can the principle of selection, which we have seen is so potent in the hands of man, apply in nature? I think we shall see that it can act most effectually. Let it be borne in mind in what an endless number of strange peculiarities our domestic productions, and, in a lesser degree, those under nature, vary; and how strong the hereditary tendency is. Under domestication, it may be truly said that the whole organisation becomes in some degree plastic. Let it be borne in mind how infinitely complex and close-fitting are the mutual relations of all organic beings to each other and to their physical conditions of life. Can it, then, be thought improbable, seeing that variations useful to man have undoubtedly occurred, that other variations useful in some way to each being in the great and complex battle of life, should sometimes occur in the course of thousands of generations? If such do occur, can we doubt (remembering that many more individuals are born than can possibly survive) that individuals having any advantage, however slight, over others, would have the best chance of surviving and of procreating their kind? On the other hand, we may feel sure that any variation in the least degree injurious would be rigidly destroyed. This preservation of favourable variations and the rejection of injurious variations, I call Natural Selection. Variations neither useful nor injurious would not be affected by natural selection, and would be left a fluctuating element, as perhaps we see in the species called polymorphic.

We shall best understand the probable course of natural selection by taking the case of a country undergoing some physical change, for instance, of climate. The proportional numbers of its inhabitants would almost immediately undergo a change, and some species might become extinct. We may conclude, from what we have seen of the intimate and complex manner in which the inhabitants of each country are bound together, that any change in the numerical proportions of some of the inhabitants, independently of the change of climate itself, would most seriously affect many of the others. If the country were open on its borders, new forms would certainly immigrate, and this also would seriously disturb the relations of some of the former inhabitants. Let it be remembered how powerful the influence of a single introduced tree or mammal has been shown to be. But in the case of an island, or of a country partly surrounded by barriers, into which new and better adapted forms could not freely enter, we should then have places in the economy of nature which would assuredly be better filled up, if some of the original inhabitants were in some manner modified; for, had the area been open to immigration, these same places would have been seized on by intruders. In such case, every slight modification, which in the course of ages chanced to arise, and which in any way favoured the individuals of any of the species, by better adapting them to their altered conditions, would tend to be preserved; and natural selection would thus have free scope for the work of improvement.

We have reason to believe, as stated in the first chapter, that a change in the conditions of life, by specially acting on the reproductive system, causes or increases variability; and in the foregoing case the conditions of life are supposed to have undergone a change, and this would manifestly be favourable to natural selection, by giving a better chance of profitable variations occurring; and unless profitable variations do occur, natural selection can do nothing. Not that, as I believe, any extreme amount of variability is necessary; as man can certainly produce great results by adding up in any given direction mere individual differences, so could Nature, but far more easily, from having incomparably longer time at her disposal. Nor do I believe that any great physical change, as of climate, or any unusual degree of isolation to check immigration, is actually necessary to produce new and unoccupied places for natural selection to fill up by modifying and improving some of the varying inhabitants. For as all the inhabitants of each country are struggling together with nicely balanced forces, extremely slight modifications in the structure or habits of one inhabitant would often give it an advantage over others; and still further modifications of the same kind would often still further increase the advantage. No country can be named in which all the native inhabitants are now so perfectly adapted to each other and to the physical conditions under which they live, that none of them could anyhow be improved; for in all countries, the natives have been so far conquered by naturalised productions, that they have allowed foreigners to take firm possession of the land. And as foreigners have thus everywhere beaten some of the natives, we may safely conclude that the natives might have been modified with advantage, so as to have better resisted such intruders.

As man can produce and certainly has produced a great result by his methodical and unconscious means of selection, what may not nature effect? Man can act only on external and visible characters: nature cares nothing for appearances, except in so far as they may be useful to any being. She can act on every internal organ, on every shade of constitutional difference, on the whole machinery of life. Man selects only for his own good; Nature only for that of the being which she tends. Every selected character is fully exercised by her; and the being is placed under well-suited conditions of life. Man keeps the natives of many climates in the same country; he seldom exercises each selected character in some peculiar and fitting manner; he feeds a long and a short beaked pigeon on the same food; he does not exercise a long-backed or long-legged quadruped in any peculiar manner; he exposes sheep with long and short wool to the same climate. He does not allow the most vigorous males to struggle for the females. He does not rigidly destroy all inferior animals, but protects during each varying season, as far as lies in his power, all his productions. He often begins his selection by some half-monstrous form; or at least by some modification prominent enough to catch his eye, or to be plainly useful to him. Under nature, the slightest difference of structure or constitution may well turn the nicely-balanced scale in the struggle for life, and so be preserved. How fleeting are the wishes and efforts of man! how short his time! and consequently how poor will his products be, compared with those accumulated by nature during whole geological periods. Can we wonder, then, that nature's productions should be far "truer" in character than man's productions; that they should be infinitely better adapted to the most complex conditions of life, and should plainly bear the stamp of far higher workmanship?

It may be said that natural selection is daily and hourly scrutinising, throughout the world, every variation, even the slightest; rejecting that which is bad, preserving and adding up all that is good; silently and insensibly working, whenever and wherever opportunity offers, at the improvement of each organic being in relation to its organic and inorganic conditions of life. We see nothing of these slow changes in progress, until the hand of time has marked the long lapse of ages, and then so imperfect is our view into long past geological ages, that we only see that the forms of life are now different from what they formerly were.

Although natural selection can act only through and for the good of each being, yet characters and structures, which we are apt to consider as of very trifling importance, may thus be acted on. When we see leaf-eating insects green, and bark-feeders mottled-grey; the alpine ptarmigan white in winter, the red-grouse the colour of heather, and the black-grouse that of peaty earth, we must believe that these tints are of service to these birds and insects in preserving them from danger. Grouse, if not destroyed at some period of their lives, would increase in countless numbers; they are known to suffer largely from birds of prey; and hawks are guided by eyesight to their prey,—so much so, that on parts of the Continent persons are warned not to keep white pigeons, as being the most liable to destruction. Hence I can see no reason to doubt that natural selection might be most effective in giving the proper colour to each kind of grouse, and in keeping that colour, when once acquired, true and constant. Nor ought we to think that the occasional destruction of an animal of any particular colour would produce little effect: we should remember how essential it is in a flock of white sheep to destroy every lamb with the faintest trace of black. In plants the down on the fruit and the colour of the flesh are considered by botanists as characters of the most trifling importance: yet we hear from an excellent horticulturist, Downing, that in the United States smooth-skinned fruits suffer far more from a beetle, a curculio, than those with down; that purple plums suffer far more from a certain disease than yellow plums; whereas another disease attacks yellow-fleshed peaches far more than those with other coloured flesh. If, with all the aids of art, these slight differences make a great difference in cultivating the several varieties, assuredly, in a state of nature, where the trees would have to struggle with other trees and with a host of enemies, such differences would effectually settle which variety, whether a smooth or downy, a yellow or purple fleshed fruit, should succeed.

In looking at many small points of difference between species, which, as far as our ignorance permits us to judge, seem to be quite unimportant, we must not forget that climate, food, &c., probably produce some slight and direct effect. It is, however, far more necessary to bear in mind that there are many unknown laws of correlation of growth, which, when one part of the organisation is modified through variation, and the modifications are accumulated by natural selection for the good of the being, will cause other modifications, often of the most unexpected nature.

As we see that those variations which under domestication appear at any particular period of life, tend to reappear in the offspring at the same period;—for instance, in the seeds of the many varieties of our culinary and agricultural plants; in the caterpillar and cocoon stages of the varieties of the silkworm; in the eggs of poultry, and in the colour of the down of their chickens; in the horns of our sheep and cattle when nearly adult;—so in a state of nature, natural selection will be enabled to act on and modify organic beings at any age, by the accumulation of profitable variations at that age, and by their inheritance at a corresponding age. If it profit a plant to have its seeds more and more widely disseminated by the wind, I can see no greater difficulty in this being effected through natural selection, than in the cotton-planter increasing and improving by selection the down in the pods on his cotton-trees. Natural selection may modify and adapt the larva of an insect to a score of contingencies, wholly different from those which concern the mature insect. These modifications will no doubt affect, through the laws of correlation, the structure of the adult; and probably in the case of those insects which live only for a few hours, and which never feed, a large part of their structure is merely the correlated result of successive changes in the structure of their larvæ. So, conversely, modifications in the adult will probably often affect the structure of the larva; but in all cases natural selection will ensure that modifications consequent on other modifications at a different period of life, shall not be in the least degree injurious: for if they became so, they would cause the extinction of the species.

Natural selection will modify the structure of the young in relation to the parent, and of the parent in relation to the young. In social animals it will adapt the structure of each individual for the benefit of the community; if each in consequence profits by the selected change. What natural selection cannot do, is to modify the structure of one species, without giving it any advantage, for the good of another species; and though statements to this effect may be found in works of natural history, I cannot find one case which will bear investigation. A structure used only once in an animal's whole life, if of high importance to it, might be modified to any extent by natural selection; for instance, the great jaws possessed by certain insects, and used exclusively for opening the cocoon—or the hard tip to the beak of nestling birds, used for breaking the egg. It has been asserted, that of the best short-beaked tumbler-pigeons more perish in the egg than are able to get out of it; so that fanciers assist in the act of hatching. Now, if nature had to make the beak of a full-grown pigeon very short for the bird's own advantage, the process of modification would be very slow, and there would be simultaneously the most rigorous selection of the young birds within the egg, which had the most powerful and hardest beaks, for all with weak beaks would inevitably perish: or, more delicate and more easily broken shells might be selected, the thickness of the shell being known to vary like every other structure.

SEXUAL SELECTION

Inasmuch as peculiarities often appear under domestication in one sex and become hereditarily attached to that sex, the same fact probably occurs under nature, and if so, natural selection will be able to modify one sex in its functional relations to the other sex, or in relation to wholly different habits of life in the two sexes, as is sometimes the case with insects. And this leads me to say a few words on what I call Sexual Selection. This depends, not on a struggle for existence, but on a struggle between the males for possession of the females; the result is not death to the unsuccessful competitor, but few or no offspring. Sexual selection is, therefore, less rigorous than natural selection. Generally, the most vigorous males, those which are best fitted for their places in nature, will leave most progeny. But in many cases, victory will depend not on general vigour, but on having special weapons, confined to the male sex. A hornless stag or spurless cock would have a poor chance of leaving offspring. Sexual selection by always allowing the victor to breed might surely give indomitable courage, length to the spur, and strength to the wing to strike in the spurred leg, as well as the brutal cockfighter, who knows well that he can improve his breed by careful selection of the best cocks. How low in the scale of nature this law of battle descends, I know not; male alligators have been described as fighting, bellowing, and whirling round, like Indians in a wardance, for the possession of the females; male salmons have been seen fighting all day long; male stag-beetles often bear wounds from the huge mandibles of other males. The war is, perhaps, severest between the males of polygamous animals, and these seem oftenest provided with special weapons. The males of carnivorous animals are already well armed; though to them and to others, special means of defence may be given through means of sexual selection, as the mane to the lion, the shoulder-pad to the boar, and the hooked jaw to the male salmon; for the shield may be as important for victory, as the sword or spear.

Amongst birds, the contest is often of a more peaceful character. All those who have attended to the subject, believe that there is the severest rivalry between the males of many species to attract by singing the females. The rock-thrush of Guiana, birds of Paradise, and some others, congregate; and successive males display their gorgeous plumage and perform strange antics before the females, which standing by as spectators, at last choose the most attractive partner. Those who have closely attended to birds in confinement well know that they often take individual preferences and dislikes: thus Sir R. Heron has described how one pied peacock was eminently attractive to all his hen birds. It may appear childish to attribute any effect to such apparently weak means: I cannot here enter on the details necessary to support this view; but if man can in a short time give elegant carriage and beauty to his bantams, according to his standard of beauty, I can see no good reason to doubt that female birds, by selecting, during thousands of generations, the most melodious or beautiful males, according to their standard of beauty, might produce a marked effect. I strongly suspect that some well-known laws with respect to the plumage of male and female birds, in comparison with the plumage of the young, can be explained on the view of plumage having been chiefly modified by sexual selection, acting when the birds have come to the breeding age or during the breeding season; the modifications thus produced being inherited at corresponding ages or seasons, either by the males alone, or by the males and females; but I have not space here to enter on this subject.

Thus it is, as I believe, that when the males and females of any animal have the same general habits of life, but differ in structure, colour, or ornament, such differences have been mainly caused by sexual selection; that is, individual males have had, in successive generations, some slight advantage over other males, in their weapons, means of defence, or charms; and have transmitted these advantages to their male offspring. Yet, I would not wish to attribute all such sexual differences to this agency: for we see peculiarities arising and becoming attached to the male sex in our domestic animals (as the wattle in male carriers, horn-like protuberances in the cocks of certain fowls, &c.), which we cannot believe to be either useful to the males in battle, or attractive to the females. We see analogous cases under nature, for instance, the tuft of hair on the breast of the turkey-cock, which can hardly be either useful or ornamental to this bird;—indeed, had the tuft appeared under domestication, it would have been called a monstrosity.

ILLUSTRATIONS OF THE ACTION OF NATURAL SELECTION

In order to make it clear how, as I believe, natural selection acts, I must beg permission to give one or two imaginary illustrations. Let us take the case of a wolf, which preys on various animals, securing some by craft, some by strength, and some by fleetness; and let us suppose that the fleetest prey, a deer for instance, had from any change in the country increased in numbers, or that other prey had decreased in numbers, during that season of the year when the wolf is hardest pressed for food. I can under such circumstances see no reason to doubt that the swiftest and slimmest wolves would have the best chance of surviving, and so be preserved or selected,—provided always that they retained strength to master their prey at this or at some other period of the year, when they might be compelled to prey on other animals. I can see no more reason to doubt this, than that man can improve the fleetness of his grey-hounds by careful and methodical selection, or by that unconscious selection which results from each man trying to keep the best dogs without any thought of modifying the breed.

Even without any change in the proportional numbers of the animals on which our wolf preyed, a cub might be born with an innate tendency to pursue certain kinds of prey. Nor can this be thought very improbable; for we often observe great differences in the natural tendencies of our domestic animals; one cat, for instance, taking to catch rats, another mice; one cat, according to Mr. St. John, bringing home winged game, another hares or rabbits, and another hunting on marshy ground and almost nightly catching woodcocks or snipes. The tendency to catch rats rather than mice is known to be inherited. Now, if any slight innate change of habit or of structure benefited an individual wolf, it would have the best chance of surviving and of leaving offspring. Some of its young would probably inherit the same habits or structure, and by the repetition of this process, a new variety might be formed which would either supplant or coexist with the parent-form of wolf. Or, again, the wolves inhabiting a mountainous district, and those frequenting the lowlands, would naturally be forced to hunt different prey; and from the continued preservation of the individuals best fitted for the two sites, two varieties might slowly be formed. These varieties would cross and blend where they met; but to this subject of intercrossing we shall soon have to return. I may add, that, according to Mr. Pierce, there are two varieties of the wolf inhabiting the Catskill Mountains in the United States, one with a light greyhound-like form, which pursues deer, and the other more bulky, with shorter legs, which more frequently attacks the shepherd's flocks.

Let us now take a more complex case. Certain plants excrete a sweet juice, apparently for the sake of eliminating something injurious from their sap: this is effected by glands at the base of the stipules in some Leguminosæ, and at the back of the leaf of the common laurel. This juice, though small in quantity, is greedily sought by insects. Let us now suppose a little sweet juice or nectar to be excreted by the inner bases of the petals of a flower. In this case insects in seeking the nectar would get dusted with pollen, and would certainly often transport the pollen from one flower to the stigma of another flower. The flowers of two distinct individuals of the same species would thus get crossed; and the act of crossing, we have good reason to believe (as will hereafter be more fully alluded to), would produce very vigorous seedlings, which consequently would have the best chance of flourishing and surviving. Some of these seedlings would probably inherit the nectar-excreting power. Those individual flowers which had the largest glands or nectaries, and which excreted most nectar, would be oftenest visited by insects, and would be oftenest crossed; and so in the long-run would gain the upper hand. Those flowers, also, which had their stamens and pistils placed, in relation to the size and habits of the particular insects which visited them, so as to favour in any degree the transportal of their pollen from flower to flower, would likewise be favoured or selected. We might have taken the case of insects visiting flowers for the sake of collecting pollen instead of nectar; and as pollen is formed for the sole object of fertilisation, its destruction appears a simple loss to the plant; yet if a little pollen were carried, at first occasionally and then habitually, by the pollen-devouring insects from flower to flower, and a cross thus effected, although nine-tenths of the pollen were destroyed, it might still be a great gain to the plant; and those individuals which produced more and more pollen, and had larger and larger anthers, would be selected.

When our plant, by this process of the continued preservation or natural selection of more and more attractive flowers, had been rendered highly attractive to insects, they would, unintentionally on their part, regularly carry pollen from flower to flower; and that they can most effectually do this, I could easily show by many striking instances. I will give only one—not as a very striking case, but as likewise illustrating one step in the separation of the sexes of plants, presently to be alluded to. Some holly-trees bear only male flowers, which have four stamens producing rather a small quantity of pollen, and a rudimentary pistil; other holly-trees bear only female flowers; these have a full-sized pistil, and four stamens with shrivelled anthers, in which not a grain of pollen can be detected. Having found a female tree exactly sixty yards from a male tree, I put the stigmas of twenty flowers, taken from different branches, under the microscope, and on all, without exception, there were pollen-grains, and on some a profusion of pollen. As the wind had set for several days from the female to the male tree, the pollen could not thus have been carried. The weather had been cold and boisterous, and therefore not favourable to bees, nevertheless every female flower which I examined had been effectually fertilised by the bees, accidentally dusted with pollen, having flown from tree to tree in search of nectar. But to return to our imaginary case: as soon as the plant had been rendered so highly attractive to insects that pollen was regularly carried from flower to flower, another process might commence. No naturalist doubts the advantage of what has been called the "physiological division of labour;" hence we may believe that it would be advantageous to a plant to produce stamens alone in one flower or on one whole plant, and pistils alone in another flower or on another plant. In plants under culture and placed under new conditions of life, sometimes the male organs and sometimes the female organs become more or less impotent; now if we suppose this to occur in ever so slight a degree under nature, then as pollen is already carried regularly from flower to flower, and as a more complete separation of the sexes of our plant would be advantageous on the principle of the division of labour, individuals with this tendency more and more increased, would be continually favoured or selected, until at last a complete separation of the sexes would be effected.

Let us now turn to the nectar-feeding insects in our imaginary case: we may suppose the plant of which we have been slowly increasing the nectar by continued selection, to be a common plant; and that certain insects depended in main part on its nectar for food. I could give many facts, showing how anxious bees are to save time; for instance, their habit of cutting holes and sucking the nectar at the bases of certain flowers, which they can, with a very little more trouble, enter by the mouth. Bearing such facts in mind, I can see no reason to doubt that an accidental deviation in the size and form of the body, or in the curvature and length of the proboscis, &c., far too slight to be appreciated by us, might profit a bee or other insect, so that an individual so characterised would be able to obtain its food more quickly, and so have a better chance of living and leaving descendants. Its descendants would probably inherit a tendency to a similar slight deviation of structure. The tubes of the corollas of the common red and incarnate clovers (Trifolium pratense and incarnatum) do not on a hasty glance appear to differ in length; yet the hive-bee can easily suck the nectar out of the incarnate clover, but not out of the common red clover, which is visited by humble-bees alone; so that whole fields of the red clover offer in vain an abundant supply of precious nectar to the hive-bee. Thus it might be a great advantage to the hive-bee to have a slightly longer or differently constructed proboscis. On the other hand, I have found by experiment that the fertility of clover greatly depends on bees visiting and moving parts of the corolla, so as to push the pollen on to the stigmatic surface. Hence, again, if humble-bees were to become rare in any country, it might be a great advantage to the red clover to have a shorter or more deeply divided tube to its corolla, so that the hive-bee could visit its flowers. Thus I can understand how a flower and a bee might slowly become, either simultaneously or one after the other, modified and adapted in the most perfect manner to each other, by the continued preservation of individuals presenting mutual and slightly favourable deviations of structure.

I am well aware that this doctrine of natural selection, exemplified in the above imaginary instances, is open to the same objections which were at first urged against Sir Charles Lyell's noble views on "the modern changes of the earth, as illustrative of geology;" but we now very seldom hear the action, for instance, of the coastwaves, called a trifling and insignificant cause, when applied to the excavation of gigantic valleys or to the formation of the longest lines of inland cliffs. Natural selection can act only by the preservation and accumulation of infinitesimally small inherited modifications, each profitable to the preserved being; and as modern geology has almost banished such views as the excavation of a great valley by a single diluvial wave, so will natural selection, if it be a true principle, banish the belief of the continued creation of new organic beings, or of any great and sudden modification in their structure.

* * *

CIRCUMSTANCES FAVOURABLE TO NATURAL SELECTION

This is an extremely intricate subject. A large amount of inheritable and diversified variability is favourable, but I believe mere individual differences suffice for the work. A large number of individuals, by giving a better chance for the appearance within any given period of profitable variations, will compensate for a lesser amount of variability in each individual, and is, I believe, an extremely important element of success. Though nature grants vast periods of time for the work of natural selection, she does not grant an indefinite period; for as all organic beings are striving, it may be said, to seize on each place in the economy of nature, if any one species does not become modified and improved in a corresponding degree with its competitors, it will soon be exterminated.

In man's methodical selection, a breeder selects for some definite object, and free intercrossing will wholly stop his work. But when many men, without intending to alter the breed, have a nearly common standard of perfection, and all try to get and breed from the best animals, much improvement and modification surely but slowly follow from this unconscious process of selection, notwithstanding a large amount of crossing with inferior animals. Thus it will be in nature; for within a confined area, with some place in its polity not so perfectly occupied as might be, natural selection will always tend to preserve all the individuals varying in the right direction, though in different degrees, so as better to fill up the unoccupied place. But if the area be large, its several districts will almost certainly present different conditions of life; and then if natural selection be modifying and improving a species in the several districts, there will be intercrossing with the other individuals of the same species on the confines of each. And in this case the effects of intercrossing can hardly be counterbalanced by natural selection always tending to modify all the individuals in each district in exactly the same manner to the conditions of each; for in a continuous area, the conditions will generally graduate away insensibly from one district to another. The intercrossing will most affect those animals which unite for each birth, which wander much, and which do not breed at a very quick rate. Hence in animals of this nature, for instance in birds, varieties will generally be confined to separated countries; and this I believe to be the case. In hermaphrodite organisms which cross only occasionally, and likewise in animals which unite for each birth, but which wander little and which can increase at a very rapid rate, a new and improved variety might be quickly formed on any one spot, and might there maintain itself in a body, so that whatever intercrossing took place would be chiefly between the individuals of the same new variety. A local variety when once thus formed might subsequently slowly spread to other districts. On the above principle, nurserymen always prefer getting seed from a large body of plants of the same variety, as the chance of intercrossing with other varieties is thus lessened.

Even in the case of slow-breeding animals, which unite for each birth, we must not overrate the effects of intercrosses in retarding natural selection; for I can bring a considerable catalogue of facts, showing that within the same area, varieties of the same animal can long remain distinct, from haunting different stations, from breeding at slightly different seasons, or from varieties of the same kind preferring to pair together.

Intercrossing plays a very important part in nature in keeping the individuals of the same species, or of the same variety, true and uniform in character. It will obviously thus act far more efficiently with those animals which unite for each birth; but I have already attempted to show that we have reason to believe that occasional intercrosses take place with all animals and with all plants. Even if these take place only at long intervals, I am convinced that the young thus produced will gain so much in vigour and fertility over the offspring from long-continued self-fertilisation, that they will have a better chance of surviving and propagating their kind; and thus, in the long run, the influence of intercrosses, even at rare intervals, will be great. If there exist organic beings which never intercross, uniformity of character can be retained amongst them, as long as their conditions of life remain the same, only through the principle of inheritance, and through natural selection destroying any which depart from the proper type; but if their conditions of life change and they undergo modification, uniformity of character can be given to their modified offspring, solely by natural selection preserving the same favourable variations.

Isolation, also, is an important element in the process of natural selection. In a confined or isolated area, if not very large, the organic and inorganic conditions of life will generally be in a great degree uniform; so that natural selection will tend to modify all the individuals of a varying species throughout the area in the same manner in relation to the same conditions. Intercrosses, also, with the individuals of the same species, which otherwise would have inhabited the surrounding and differently circumstanced districts, will be prevented. But isolation probably acts more efficiently in checking the immigration of better adapted organisms, after any physical change, such as of climate or elevation of the land, &c.; and thus new places in the natural economy of the country are left open for the old inhabitants to struggle for, and become adapted to, through modifications in their structure and constitution. Lastly, isolation, by checking immigration and consequently competition, will give time for any new variety to be slowly improved; and this may sometimes be of importance in the production of new species. If, however, an isolated area be very small, either from being surrounded by barriers, or from having very peculiar physical conditions, the total number of the individuals supported on it will necessarily be very small; and fewness of individuals will greatly retard the production of new species through natural selection, by decreasing the chance of the appearance of favourable variations.

If we turn to nature to test the truth of these remarks, and look at any small isolated area, such as an oceanic island, although the total number of the species inhabiting it, will be found to be small, as we shall see in our chapter on geographical distribution; yet of these species a very large proportion are endemic,—that is, have been produced there, and nowhere else. Hence an oceanic island at first sight seems to have been highly favourable for the production of new species. But we may thus greatly deceive ourselves, for to ascertain whether a small isolated area, or a large open area like a continent, has been most favourable for the production of new organic forms, we ought to make the comparison within equal times; and this we are incapable of doing.

Although I do not doubt that isolation is of considerable importance in the production of new species, on the whole I am inclined to believe that largeness of area is of more importance, more especially in the production of species, which will prove capable of enduring for a long period, and of spreading widely. Throughout a great and open area, not only will there be a better chance of favourable variations arising from the large number of individuals of the same species there supported, but the conditions of life are infinitely complex from the large number of already existing species; and if some of these many species become modified and improved, others will have to be improved in a corresponding degree or they will be exterminated. Each new form, also, as soon as it has been much improved, will be able to spread over the open and continuous area, and will thus come into competition with many others. Hence more new places will be formed, and the competition to fill them will be more severe, on a large than on a small and isolated area. Moreover, great areas, though now continuous, owing to oscillations of level, will often have recently existed in a broken condition, so that the good effects of isolation will generally, to a certain extent, have concurred. Finally, I conclude that, although small isolated areas probably have been in some respects highly favourable for the production of new species, yet that the course of modification will generally have been more rapid on large areas; and what is more important, that the new forms produced on large areas, which already have been victorious over many competitors, will be those that will spread most widely, will give rise to most new varieties and species, and will thus play an important part in the changing history of the organic world.

We can, perhaps, on these views, understand some facts which will be again alluded to in our chapter on geographical distribution; for instance, that the productions of the smaller continent of Australia have formerly yielded, and apparently are now yielding, before those of the larger Europæo-Asiatic area. Thus, also, it is that continental productions have everywhere become so largely naturalised on islands. On a small island, the race for life will have been less severe, and there will have been less modification and less extermination. Hence, perhaps, it comes that the flora of Madeira, according to Oswald Heer, resembles the extinct tertiary flora of Europe. All fresh-water basins, taken together, make a small area compared with that of the sea or of the land; and, consequently, the competition between fresh-water productions will have been less severe than elsewhere; new forms will have been more slowly formed, and old forms more slowly exterminated. And it is in fresh water that we find seven genera of Ganoid fishes, remnants of a once preponderant order: and in fresh water we find some of the most anomalous forms now known in the world, as the Ornithorhynchus and Lepidosiren, which, like fossils, connect to a certain extent orders now widely separated in the natural scale. These anomalous forms may almost be called living fossils; they have endured to the present day, from having inhabited a confined area, and from having thus been exposed to less severe competition.

To sum up the circumstances favourable and unfavourable to natural selection, as far as the extreme intricacy of the subject permits. I conclude, looking to the future, that for terrestrial productions a large continental area, which will probably undergo many oscillations of level, and which consequently will exist for long periods in a broken condition, will be the most favourable for the production of many new forms of life, likely to endure long and to spread widely. For the area will first have existed as a continent, and the inhabitants, at this period numerous in individuals and kinds, will have been subjected to very severe competition. When converted by subsidence into large separate islands, there will still exist many individuals of the same species on each island: intercrossing on the confines of the range of each species will thus be checked: after physical changes of any kind, immigration will be prevented, so that new places in the polity of each island will have to be filled up by modifications of the old inhabitants; and time will be allowed for the varieties in each to become well modified and perfected. When, by renewed elevation, the islands shall be reconverted into a continental area, there will again be severe competition: the most favoured or improved varieties will be enabled to spread: there will be much extinction of the less improved forms, and the relative proportional numbers of the various inhabitants of the renewed continent will again be changed; and again there will be a fair field for natural selection to improve still further the inhabitants, and thus produce new species.

That natural selection will always act with extreme slowness, I fully admit. Its action depends on there being places in the polity of nature, which can be better occupied by some of the inhabitants of the country undergoing modification of some kind. The existence of such places will often depend on physical changes, which are generally very slow, and on the immigration of better adapted forms having been checked. But the action of natural selection will probably still oftener depend on some of the inhabitants becoming slowly modified; the mutual relations of many of the other inhabitants being thus disturbed. Nothing can be effected, unless favourable variations occur, and variation itself is apparently always a very slow process. The process will often be greatly retarded by free intercrossing. Many will exclaim that these several causes are amply sufficient wholly to stop the action of natural selection. I do not believe so. On the other hand, I do believe that natural selection will always act very slowly, often only at long intervals of time, and generally on only a very few of the inhabitants of the same region at the same time. I further believe, that this very slow, intermittent action of natural selection accords perfectly well with what geology tells us of the rate and manner at which the inhabitants of this world have changed.

Slow though the process of selection may be, if feeble man can do much by his powers of artificial selection, I can see no limit to the amount of change, to the beauty and infinite complexity of the coadaptations between all organic beings, one with another and with their physical conditions of life, which may be effected in the long course of time by nature's power of selection.

* * *

DIVERGENCE OF CHARACTER

The principle, which I have designated by this term, is of high importance on my theory, and explains, as I believe, several important facts. In the first place, varieties, even strongly-marked ones, though having somewhat of the character of species—as is shown by the hopeless doubts in many cases how to rank them—yet certainly differ from each other far less than do good and distinct species. Nevertheless, according to my view, varieties are species in the process of formation, or are, as I have called them, incipient species. How, then, does the lesser difference between varieties become augmented into the greater difference between species? That this does habitually happen, we must infer from most of the innumerable species throughout nature presenting well-marked differences; whereas varieties, the supposed prototypes and parents of future well-marked species, present slight and ill-defined differences. Mere chance, as we may call it, might cause one variety to differ in some character from its parents, and the offspring of this variety again to differ from its parent in the very same character and in a greater degree; but this alone would never account for so habitual and large an amount of difference as that between varieties of the same species and species of the same genus.

As has always been my practice, let us seek light on this head from our domestic productions. We shall here find something analogous. A fancier is struck by a pigeon having a slightly shorter beak; another fancier is struck by a pigeon having a rather longer beak; and on the acknowledged principle that "fanciers do not and will not admire a medium standard, but like extremes," they both go on (as has actually occurred with tumbler-pigeons) choosing and breeding from birds with longer and longer beaks, or with shorter and shorter beaks. Again, we may suppose that at an early period one man preferred swifter horses; another stronger and more bulky horses. The early differences would be very slight; in the course of time, from the continued selection of swifter horses by some breeders, and of stronger ones by others, the differences would become greater, and would be noted as forming two sub-breeds; finally, after the lapse of centuries, the sub-breeds would become converted into two well-established and distinct breeds. As the differences slowly become greater, the inferior animals with intermediate characters, being neither very swift nor very strong, will have been neglected, and will have tended to disappear. Here, then, we see in man's productions the action of what may be called the principle of divergence, causing differences, at first barely appreciable, steadily to increase, and the breeds to diverge in character both from each other and from their common parent.

But how, it may be asked, can any analogous principle apply in nature? I believe it can and does apply most efficiently, from the simple circumstance that the more diversified the descendants from any one species become in structure, constitution, and habits, by so much will they be better enabled to seize on many and widely diversified places in the polity of nature, and so be enabled to increase in numbers.

We can clearly see this in the case of animals with simple habits. Take the case of a carnivorous quadruped, of which the number that can be supported in any country has long ago arrived at its full average. If its natural powers of increase be allowed to act, it can succeed in increasing (the country not undergoing any change in its conditions) only by its varying descendants seizing on places at present occupied by other animals: some of them, for instance, being enabled to feed on new kinds of prey, either dead or alive; some inhabiting new stations, climbing trees, frequenting water, and some perhaps becoming less carnivorous. The more diversified in habits and structure the descendants of our carnivorous animal became, the more places they would be enabled to occupy. What applies to one animal will apply throughout all time to all animals —that is, if they vary—for otherwise natural selection can do nothing. So it will be with plants. It has been experimentally proved, that if a plot of ground be sown with one species of grass, and a similar plot be sown with several distinct genera of grasses, a greater number of plants and a greater weight of dry herbage can thus be raised. The same has been found to hold good when first one variety and then several mixed varieties of wheat have been sown on equal spaces of ground. Hence, if any one species of grass were to go on varying, and those varieties were continually selected which differed from each other in at all the same manner as distinct species and genera of grasses differ from each other, a greater number of individual plants of this species of grass, including its modified descendants, would succeed in living on the same piece of ground. And we well know that each species and each variety of grass is annually sowing almost countless seeds; and thus, as it may be said, is striving its utmost to increase its numbers. Consequently, I cannot doubt that in the course of many thousands of generations, the most distinct varieties of any one species of grass would always have the best chance of succeeding and of increasing in numbers, and thus of supplanting the less distinct varieties; and varieties, when rendered very distinct from each other, take the rank of species.

The truth of the principle, that the greatest amount of life can be supported by great diversification of structure, is seen under many natural circumstances. In an extremely small area, especially if freely open to immigration, and where the contest between individual and individual must be severe, we always find great diversity in its inhabitants. For instance, I found that a piece of turf, three feet by four in size, which had been exposed for many years to exactly the same conditions, supported twenty species of plants, and these belonged to eighteen genera and to eight orders, which shows how much these plants differed from each other. So it is with the plants and insects on small and uniform islets; and so in small ponds of fresh water. Farmers find that they can raise most food by a rotation of plants belonging to the most different orders: nature follows what may be called a simultaneous rotation. Most of the animals and plants which live close round any small piece of ground, could live on it (supposing it not to be in any way peculiar in its nature), and may be said to be striving to the utmost to live there; but, it is seen, that where they come into the closest competition with each other, the advantages of diversification of structure, with the accompanying differences of habit and constitution, determine that the inhabitants, which thus jostle each other most closely, shall, as a general rule, belong to what we call different genera and orders.

The same principle is seen in the naturalisation of plants through man's agency in foreign lands. It might have been expected that the plants which have succeeded in becoming naturalised in any land would generally have been closely allied to the indigenes; for these are commonly looked at as specially created and adapted for their own country. It might, also, perhaps have been expected that naturalised plants would have belonged to a few groups more especially adapted to certain stations in their new homes. But the case is very different; and Alph. De Candolle has well remarked in his great and admirable work, that floras gain by naturalisation, proportionally with the number of the native genera and species, far more in new genera than in new species. To give a single instance: in the last edition of Dr. Asa Gray's 'Manual of the Flora of the Northern United States,' 260 naturalised plants are enumerated, and these belong to 162 genera. We thus see that these naturalised plants are of a highly diversified nature. They differ, moreover, to a large extent from the indigenes, for out of the 162 genera, no less than 100 genera are not there indigenous, and thus a large proportional addition is made to the genera of these States.

By considering the nature of the plants or animals which have struggled successfully with the indigenes of any country, and have there become naturalised, we can gain some crude idea in what manner some of the natives would have had to be modified, in order to have gained an advantage over the other natives; and we may, I think, at least safely infer that diversification of structure, amounting to new generic differences, would have been profitable to them.

The advantage of diversification in the inhabitants of the same region is, in fact, the same as that of the physiological division of labour in the organs of the same individual body—a subject so well elucidated by Milne Edwards. No physiologist doubts that a stomach by being adapted to digest vegetable matter alone, or flesh alone, draws most nutriment from these substances. So in the general economy of any land, the more widely and perfectly the animals and plants are diversified for different habits of life, so will a greater number of individuals be capable of there supporting themselves. A set of animals, with their organisation but little diversified, could hardly compete with a set more perfectly diversified in structure. It may be doubted, for instance, whether the Australian marsupials, which are divided into groups differing but little from each other, and feebly representing, as Mr. Waterhouse and others have remarked, our carnivorous, ruminant, and rodent mammals, could successfully compete with these well-pronounced orders. In the Australian mammals, we see the process of diversification in an early and incomplete stage of development.

After the foregoing discussion, which ought to have been much amplified, we may, I think, assume that the modified descendants of any one species will succeed by so much the better as they become more diversified in structure, and are thus enabled to encroach on places occupied by other beings. Now let us see how this principle of great benefit being derived from divergence of character, combined with the principles of natural selection and of extinction, will tend to act.

The accompanying diagram will aid us in understanding this rather perplexing subject. Let A to L represent the species of a genus large in its own country; these species are supposed to resemble each other in unequal degrees, as is so generally the case in nature, and as is represented in the diagram by the letters standing at unequal distances. I have said a large genus, because we have seen in the second chapter, that on an average more of the species of large genera vary than of small genera; and the varying species of the large genera present a greater number of varieties. We have, also, seen that the species, which are the commonest and the most widely-diffused, vary more than rare species with restricted ranges. Let (A) be a common, widely-diffused, and varying species, belonging to a genus large in its own country. The little fan of diverging dotted lines of unequal lengths proceeding from (A), may represent its varying offspring. The variations are supposed to be extremely slight, but of the most diversified nature; they are not supposed all to appear simultaneously, but often after long intervals of time; nor are they all supposed to endure for equal periods. Only those variations which are in some way profitable will be preserved or naturally selected. And here the importance of the principle of benefit being derived from divergence of character comes in; for this will generally lead to the most different or divergent variations (represented by the outer dotted lines) being preserved and accumulated by natural selection. When a dotted line reaches one of the horizontal lines, and is there marked by a small numbered letter, a sufficient amount of variation is supposed to have been accumulated to have formed a fairly well-marked variety, such as would be thought worthy of record in a systematic work.

The intervals between the horizontal lines in the diagram, may represent each a thousand generations; but it would have been better if each had represented ten thousand generations. After a thousand generations, species (A) is supposed to have produced two fairly well-marked varieties, namely a^1 and m^1. These two varieties will generally continue to be exposed to the same conditions which

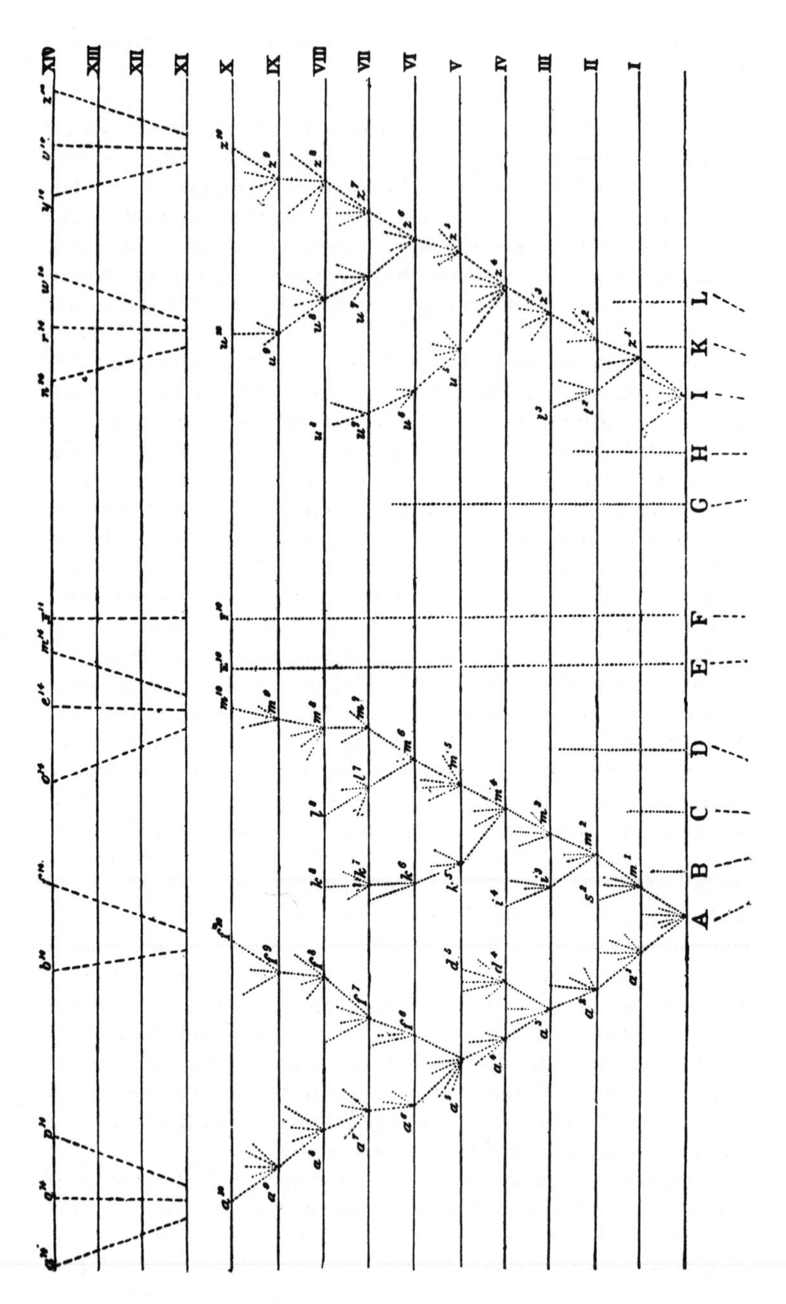
A
B
C
D
E
F
G
H
I
K
L
I
II
III
IV
V
VI
VII
VIII
IX
X
XI
XII
XIII
XIV

made their parents variable, and the tendency to variability is in itself hereditary, consequently they will tend to vary, and generally to vary in nearly the same manner as their parents varied. Moreover, these two varieties, being only slightly modified forms, will tend to inherit those advantages which made their common parent (A) more numerous than most of the other inhabitants of the same country; they will likewise partake of those more general advantages which made the genus to which the parent-species belonged, a large genus in its own country. And these circumstances we know to be favourable to the production of new varieties.

If, then, these two varieties be variable, the most divergent of their variations will generally be preserved during the next thousand generations. And after this interval, variety a^1 is supposed in the diagram to have produced variety a^2, which will, owing to the principle of divergence, differ more from (A) than did variety a^1. Variety m^1 is supposed to have produced two varieties, namely m^2 and s^3, differing from each other, and more considerably from their common parent (A). We may continue the process by similar steps for any length of time; some of the varities, after each thousand generations, producing only a single variety, but in a more and more modified condition, some producing two or three varieties, and some failing to produce any. Thus the varieties or modified descendants, proceeding from the common parent (A), will generally go on increasing in number and diverging in character. In the diagram the process is represented up to the ten-thousandth generation, and under a condensed and simplified form up to the fourteen-thousandth generation.

But I must here remark that I do not suppose that the process ever goes on so regularly as is represented in the diagram, though in itself made somewhat irregular. I am far from thinking that the most divergent varieties will invariably prevail and multiply: a medium form may often long endure, and may or may not produce more than one modified descendant; for natural selection will always act according to the nature of the places which are either unoccupied or not perfectly occupied by other beings; and this will depend on infinitely complex relations. But as a general rule, the more diversified in structure the descendants from any one species can be rendered, the more places they will be enabled to seize on, and the more their modified progeny will be increased. In our diagram the line of succession is broken at regular intervals by small numbered letters marking the successive forms which have become sufficiently distinct to be recorded as varieties. But these breaks are imaginary, and might have been inserted anywhere, after intervals long enough to have allowed the accumulation of a considerable amount of divergent variation.

As all the modified descendants from a common and widely-diffused species, belonging to a large genus, will tend to partake of the same advantages which made their parent successful in life, they will generally go on multiplying in number as well as diverging in character: this is represented in the diagram by the several divergent branches proceeding from (A). The modified offspring from the later and more highly improved branches in the lines of descent, will, it is probable, often take the place of, and so destroy, the earlier and less improved branches: this is represented in the diagram by some of the lower branches not reaching to the upper horizontal lines. In some cases I do not doubt that the process of modification will be confined to a single line of descent, and the number of the descendants will not be increased; although the amount of divergent modification may have been increased in the successive generations. This case would be represented in the diagram, if all the lines proceeding from (A) were removed, excepting that from a^1 to a^{10}. In the same way, for instance, the English race-horse and English pointer have apparently both gone on slowly diverging in character from their original stocks, without either having given off any fresh branches or races.

After ten thousand generations, species (A) is supposed to have produced three forms, a^{10}, f^{10}, and m^{10}, which, from having diverged in character during the successive generations, will have come to differ largely, but perhaps unequally, from each other and from their common parent. If we suppose the amount of change between each horizontal line in our diagram to be excessively small, these three forms may still be only well-marked varieties; or they may have arrived at the doubtful category of sub-species; but we have only to suppose the steps in the process of modification to be more numerous or greater in amount, to convert these three forms into well-defined species: thus the diagram illustrates the steps by which the small differences distinguishing varieties are increased into the larger differences distinguishing species. By continuing the same process for a greater number of generations (as shown in the diagram in a condensed and simplified manner), we get eight species, marked by the letters between a^{14} and m^{14}, all descended from (A). Thus, as I believe, species are multiplied and genera are formed.

In a large genus it is probable that more than one species would vary. In the diagram I have assumed that a second species (I) has produced, by analogous steps, after ten thousand generations, either two well-marked varieties (w^{10} and z^{10}) or two species, according to the amount of change supposed to be represented between the horizontal lines. After fourteen thousand generations, six new species, marked by the letters n^{14} to z^{14}, are supposed to have been produced. In each genus, the species, which are already extremely different in character, will generally tend to produce the greatest number of modified descendants; for these will have the best chance of filling new and widely different places in the polity of nature: hence in the diagram I have chosen the extreme species (A), and the nearly extreme species (I), as those which have largely varied, and have given rise to new varieties and species. The other nine species (marked by capital letters) of our original genus, may for a long period continue transmitting unaltered descendants; and this is shown in the diagram by the dotted lines not prolonged far upwards from want of space.

But during the process of modification, represented in the diagram, another of our principles, namely that of extinction, will have played an important part. As in each fully stocked country natural selection necessarily acts by the selected form having some advantage in the struggle for life over other forms, there will be a constant tendency in the improved descendants of any one species to supplant and exterminate in each stage of descent their predecessors and their original parent. For it should be remembered that the competition will generally be most severe between those forms which are most nearly related to each other in habits, constitution, and structure. Hence all the intermediate forms between the earlier and later states, that is between the less and more improved state of a species, as well as the original parent-species itself, will generally tend to become extinct. So it probably will be with many whole collateral lines of descent, which will be conquered by later and improved lines of descent. If, however, the modified offspring of a species get into some distinct country, or become quickly adapted to some quite new station, in which child and parent do not come into competition, both may continue to exist.

If then our diagram be assumed to represent a considerable amount of modification, species (A) and all the earlier varieties will have become extinct, having been replaced by eight new species (a^{14} to m^{14}); and (I) will have been replaced by six (n^{14} to z^{14}) new species.

But we may go further than this. The original species of our genus were supposed to resemble each other in unequal degrees, as is so generally the case in nature; species (A) being more nearly related to B, C, and D, than to the other species; and species (I) more to G, H, K, L, than to the others. These two species (A) and (I), were also supposed to be very common and widely diffused species, so that they must originally have had some advantage over most of the other species of the genus. Their modified descendants, fourteen in number at the fourteen-thousandth generation, will probably have inherited some of the same advantages: they have also been modified and improved in a diversified manner at each stage of descent, so as to have become adapted to many related places in the natural economy of their country. It seems, therefore, to me extremely probable that they will have taken the places of, and thus exterminated, not only their parents (A) and (I), but likewise some of the original species which were most nearly related to their parents. Hence very few of the original species will have transmitted offspring to the fourteen-thousandth generation. We may suppose that only one (F), of the two species which were least closely related to the other nine original species, has transmitted descendants to this late stage of descent.

The new species in our diagram descended from the original eleven species, will now be fifteen in number. Owing to the divergent tendency of natural selection, the extreme amount of difference in character between species a^{14} and z^{14} will be much greater than that between the most different of the original eleven species. The new species, moreover, will be allied to each other in a widely different manner. Of the eight descendants from (A) the three marked a^{14}, q^{14}, p^{14}, will be nearly related from having recently branched off from a^{10}; b^{14} and f^{14}, from having diverged at an earlier period from a^5, will be in some degree distinct from the three first-named species; and lastly, o^{14}, e^{14}, and m^{14}, will be nearly related one to the other, but from having diverged at the first commencement of the process of modification, will be widely different from the other five species, and may constitute a sub-genus or even a distinct genus.

The six descendants from (I) will form two sub-genera or even genera. But as the original species (I) differed largely from (A), standing nearly at the extreme points of the original genus, the six descendants from (I) will, owing to inheritance, differ considerably from the eight descendants from (A); the two groups, moreover, are supposed to have gone on diverging in different directions. The intermediate species, also (and this is a very important consideration), which connected the original species (A) and (I), have all become, excepting (F), extinct, and have left no descendants. Hence the six new species descended from (I), and the eight descended from (A), will have to be ranked as very distinct genera, or even as distinct sub-families.

Thus it is, as I believe, that two or more genera are produced by descent, with modification, from two or more species of the same genus. And the two or more parent-species are supposed to have descended from some one species of an earlier genus. In our diagram, this is indicated by the broken lines, beneath the capital letters, converging in sub-branches downwards towards a single point; this point representing a single species, the supposed single parent of our several new sub-genera and genera.

It is worth while to reflect for a moment on the character of the new species F^{14}, which is supposed not to have diverged much in character, but to have retained the form of (F), either unaltered or altered only in a slight degree. In this case, its affinities to the other fourteen new species will be of a curious and circuitous nature. Having descended from a form which stood between the two parent-species (A) and (I), now supposed to be extinct and unknown, it will be in some degree intermediate in character between the two groups descended from these species. But as these two groups have gone on diverging in character from the type of their parents, the new species (F^{14}) will not be directly intermediate between them, but rather between types of the two groups; and every naturalist will be able to bring some such case before his mind.

In the diagram, each horizontal line has hitherto been supposed to represent a thousand generations, but each may represent a million or hundred million generations, and likewise a section of the successive strata of the earth's crust including extinct remains. We shall, when we come to our chapter on Geology, have to refer again to this subject, and I think we shall then see that the diagram throws light on the affinities of extinct beings, which, though generally belonging to the same orders, or families, or genera, with those now living, yet are often, in some degree, intermediate in character between existing groups; and we can understand this fact, for the extinct species lived at very ancient epochs when the branching lines of descent had diverged less.

I see no reason to limit the process of modification, as now explained, to the formation of genera alone. If, in our diagram, we suppose the amount of change represented by each successive group of diverging dotted lines to be very great, the forms marked a^{14} to p^{14}, those marked b^{14} and f^{14}, and those marked o^{14} to m^{14}, will form three very distinct genera. We shall also have two very distinct genera descended from (I); and as these latter two genera, both from continued divergence of character and from inheritance from a different parent, will differ widely from the three genera descended from (A), the two little groups of genera will form two distinct families, or even orders, according to the amount of divergent modification supposed to be represented in the diagram. And the two new families, or orders, will have descended from two species of the original genus; and these two species are supposed to have descended from one species of a still more ancient and unknown genus.

We have seen that in each country it is the species of the larger genera which oftenest present varieties or incipient species. This, indeed, might have been expected; for as natural selection acts through one form having some advantage over other forms in the struggle for existence, it will chiefly act on those which already have some advantage; and the largeness of any group shows that its species have inherited from a common ancestor some advantage in common. Hence, the struggle for the production of new and modified descendants, will mainly lie between the larger groups, which are all trying to increase in number. One large group will slowly conquer another large group, reduce its numbers, and thus lessen its chance of further variation and improvement. Within the same large group, the later and more highly perfected sub-groups, from branching out and seizing on many new places in the polity of Nature, will constantly tend to supplant and destroy the earlier and less improved sub-groups. Small and broken groups and sub-groups will finally tend to disappear. Looking to the future, we can predict that the groups of organic beings which are now large and triumphant, and which are least broken up, that is, which as yet have suffered least extinction, will for a long period continue to increase. But which groups will ultimately prevail, no man can predict; for we well know that many groups, formerly most extensively developed, have now become extinct. Looking still more remotely to the future, we may predict that, owing to the continued and steady increase of the larger groups, a multitude of smaller groups will become utterly extinct, and leave no modified descendants; and consequently that of the species living at any one period, extremely few will transmit descendants to a remote futurity. I shall have to return to this subject in the chapter on Classification, but I may add that on this view of extremely few of the more ancient species having transmitted descendants, and on the view of all the descendants of the same species making a class, we can understand how it is that there exist but very few classes in each main division of the animal and vegetable kingdoms. Although extremely few of the most ancient species may now have living and modified descendants, yet at the most remote geological period, the earth may have been as well peopled with many species of many genera, families, orders, and classes, as at the present day.

SUMMARY OF CHAPTER

If during the long course of ages and under varying conditions of life, organic beings vary at all in the several parts of their organisation, and I think this cannot be disputed; if there be, owing to the high geometrical powers of increase of each species, at some age, season, or year, a severe struggle for life, and this certainly cannot be disputed; then, considering the infinite complexity of the relations of all organic beings to each other and to their conditions of existence, causing an infinite diversity in structure, constitution, and habits, to be advantageous to them, I think it would me a most extraordinary fact if no variation ever had occurred useful to each being's own welfare, in the same way as so many variations have occurred useful to man. But if variations useful to any organic being do occur, assuredly individuals thus characterised will have the best chance of being preserved in the struggle for life; and from the strong principle of inheritance they will tend to produce offspring similarly characterised. This principle of preservation, I have called, for the sake of brevity, Natural Selection. Natural selection, on the principle of qualities being inherited at corresponding ages, can modify the egg, seed, or young, as easily as the adult. Amongst many animals, sexual selection will give its aid to ordinary selection, by assuring to the most vigorous and best adapted males the greatest number of offspring. Sexual selection will also give characters useful to the males alone, in their struggles with other males.

Whether natural selection has really thus acted in nature, in modifying and adapting the various forms of life to their several conditions and stations, must be judged of by the general tenour and balance of evidence given in the following chapters. But we already see how it entails extinction; and how largely extinction has acted in the world's history, geology plainly declares. Natural selection, also, leads to divergence of character; for more living beings can be supported on the same area the more they diverge in structure, habits, and constitution, of which we see proof by looking at the inhabitants of any small spot or at naturalised productions. Therefore during the modification of the descendants of any one species, and during the incessant struggle of all species to increase in numbers, the more diversified these descendants become, the better will be their chance of succeeding in the battle of life. Thus the small differences distinguishing varieties of the same species, will steadily tend to increase till they come to equal the greater differences between species of the same genus, or even of distinct genera.

We have seen that it is the common, the widely-diffused, and widely-ranging species, belonging to the larger genera, which vary most; and these will tend to transmit to their modified offspring that superiority which now makes them dominant in their own countries. Natural selection, as has just been remarked, leads to divergence of character and to much extinction of the less improved and intermediate forms of life. On these principles, I believe, the nature of the affinities of all organic beings may be explained. It is a truly wonderful fact—the wonder of which we are apt to overlook from familiarity—that all animals and all plants throughout all time and space should be related to each other in group subordinate to group, in the manner which we everywhere behold—namely, varieties of the same species most closely related together, species of the same genus less closely and unequally related together, forming sections and subgenera, species of distinct genera much less closely related, and genera related in different degrees, forming subfamilies, families, orders, subclasses, and classes. The several subordinate groups in any class cannot be ranked in a single file, but seem rather to be clustered round points, and these round other points, and so on in almost endless cycles. On the view that each species has been independently created, I can see no explanation of this great fact in the classification of all organic beings; but, to the best of my judgment, it is explained through inheritance and the complex action of natural selection, entailing extinction and divergence of character, as we have seen illustrated in the diagram.

The affinities of all the beings of the same class have sometimes been represented by a great tree. I believe this simile largely speaks the truth. The green and budding twigs may represent existing species; and those produced during each former year may represent the long succession of extinct species. At each period of growth all the growing twigs have tried to branch out on all sides, and to overtop and kill the surrounding twigs and branches, in the same manner as species and groups of species have tried to overmaster other species in the great battle for life. The limbs divided into great branches, and these into lesser and lesser branches, were themselves once, when the tree was small, budding twigs; and this connexion of the former and present buds by ramifying branches may well represent the classification of all extinct and living species in groups subordinate to groups. Of the many twigs which flourished when the tree was a mere bush, only two or three, now grown into great branches, yet survive and bear all the other branches; so with the species which lived during long-past geological periods, very few now have living and modified descendants. From the first growth of the tree, many a limb and branch has decayed and dropped off; and these lost branches of various sizes may represent those whole orders, families, and genera which have now no living representatives, and which are known to us only from having been found in a fossil state. As we here and there see a thin straggling branch springing from a fork low down in a tree, and which by some chance has been favoured and is still alive on its summit, so we occasionally see an animal like the Ornithorhynchus or Lepidosiren, which in some small degree connects by its affinities two large branches of life, and which has apparently been saved from fatal competition by having inhabited a protected station. As buds give rise by growth to fresh buds, and these, if vigorous, branch out and overtop on all sides many a feebler branch, so by generation I believe it has been with the great Tree of Life, which fills with its dead and broken branches the crust of the earth, and covers the surface with its ever branching and beautiful ramifications.

Chapter VI. Difficulties on Theory

Difficulties on the theory of descent with modification—Transitions—Absence or rarity of transitional varieties—Transitions in habits of life—Diversified habits in the same species—Species with habits widely different from those of their allies—Organs of extreme perfection—Means of transition—Cases of difficulty—Natura non facit saltum—Organs of small importance—Organs not in all cases absolutely perfect—The law of Unity of Type and of the Conditions of Existence embraced by the theory of Natural Selection.

Long before having arrived at this part of my work, a crowd of difficulties will have occurred to the reader. Some of them are so grave that to this day I can never reflect on them without being staggered; but, to the best of my judgment, the greater number are only apparent, and those that are real are not, I think, fatal to my theory.

These difficulties and objections may be classed under the following heads:—Firstly, why, if species have descended from other species by insensibly fine gradations, do we not everywhere see innumerable transitional forms? Why is not all nature in confusion instead of the species being, as we see them, well defined?

Secondly, is it possible that an animal having, for instance, the structure and habits of a bat, could have been formed by the modification of some animal with wholly different habits? Can we believe that natural selection could produce, on the one hand, organs of trifling importance, such as the tail of a giraffe, which serves as a fly-flapper, and, on the other hand, organs of such wonderful structure, as the eye, of which we hardly as yet fully understand the inimitable perfection?

Thirdly, can instincts be acquired and modified through natural selection? What shall we say to so marvellous an instinct as that which leads the bee to make cells, which have practically anticipated the discoveries of profound mathematicians?

Fourthly, how can we account for species, when crossed, being sterile and producing sterile offspring, whereas, when varieties are crossed, their fertility is unimpaired?

The two first heads shall be here discussed—Instinct and Hybridism in separate chapters.

ON THE ABSENCE OR RARITY OF TRANSITIONAL VARIETIES

As natural selection acts solely by the preservation of profitable modifications, each new form will tend in a fully-stocked country to take the place of, and finally to exterminate, its own less improved parent or other less-favoured forms with which it comes into competition. Thus extinction and natural selection will, as we have seen, go hand in hand. Hence, if we look at each species as descended from some other unknown form, both the parent and all the transitional varieties will generally have been exterminated by the very process of formation and perfection of the new form.

But, as by this theory innumerable transitional forms must have existed, why do we not find them embedded in countless numbers in the crust of the earth? It will be much more convenient to discuss this question in the chapter on the Imperfection of the geological record; and I will here only state that I believe the answer mainly lies in the record being incomparably less perfect than is generally supposed; the imperfection of the record being chiefly due to organic beings not inhabiting profound depths of the sea, and to their remains being embedded and preserved to a future age only in masses of sediment sufficiently thick and extensive to withstand an enormous amount of future degradation; and such fossiliferous masses can be accumulated only where much sediment is deposited on the shallow bed of the sea, whilst it slowly subsides. These contingencies will concur only rarely, and after enormously long intervals. Whilst the bed of the sea is stationary or is rising, or when very little sediment is being deposited, there will be blanks in our geological history. The crust of the earth is a vast museum; but the natural collections have been made only at intervals of time immensely remote.

But it may be urged that when several closely-allied species inhabit the same territory we surely ought to find at the present time many transitional forms. Let us take a simple case: in travelling from north to south over a continent, we generally meet at successive intervals with closely allied or representative species, evidently filling nearly the same place in the natural economy of the land. These representative species often meet and interlock; and as the one becomes rarer and rarer, the other becomes more and more frequent, till the one replaces the other. But if we compare these species where they intermingle, they are generally as absolutely distinct from each other in every detail of structure as are specimens taken from the metropolis inhabited by each. By my theory these allied species have descended from a common parent; and during the process of modification, each has become adapted to the conditions of life of its own region, and has supplanted and exterminated its original parent and all the transitional varieties between its past and present states. Hence we ought not to expect at the present time to meet with numerous transitional varieties in each region, though they must have existed there, and may be embedded there in a fossil condition. But in the intermediate region, having intermediate conditions of life, why do we not now find closely-linking intermediate varieties? This difficulty for a long time quite confounded me. But I think it can be in large part explained.

In the first place we should be extremely cautious in inferring, because an area is now continuous, that it has been continuous during a long period. Geology would lead us to believe that almost every continent has been broken up into islands even during the later tertiary periods; and in such islands distinct species might have been separately formed without the possibility of intermediate varieties existing in the intermediate zones. By changes in the form of the land and of climate, marine areas now continuous must often have existed within recent times in a far less continuous and uniform condition than at present. But I will pass over this way of escaping from the difficulty; for I believe that many perfectly defined species have been formed on strictly continuous areas; though I do not doubt that the formerly broken condition of areas now continuous has played an important part in the formation of new species, more especially with freely-crossing and wandering animals.

In looking at species as they are now distributed over a wide area, we generally find them tolerably numerous over a large territory, then becoming somewhat abruptly rarer and rarer on the confines, and finally disappearing. Hence the neutral territory between two representative species is generally narrow in comparison with the territory proper to each. We see the same fact in ascending mountains, and sometimes it is quite remarkable how abruptly, as Alph. De Candolle has observed, a common alpine species disappears. The same fact has been noticed by Forbes in sounding the depths of the sea with the dredge. To those who look at climate and the physical conditions of life as the all-important elements of distribution, these facts ought to cause surprise, as climate and height or depth graduate away insensibly. But when we bear in mind that almost every species, even in its metropolis, would increase immensely in numbers, were it not for other competing species; that nearly all either prey on or serve as prey for others; in short, that each organic being is either directly or indirectly related in the most important manner to other organic beings, we must see that the range of the inhabitants of any country by no means exclusively depends on insensibly changing physical conditions, but in large part on the presence of other species, on which it depends, or by which it is destroyed, or with which it comes into competition; and as these species are already defined objects (however they may have become so), not blending one into another by insensible gradations, the range of any one species, depending as it does on the range of others, will tend to be sharply defined. Moreover, each species on the confines of its range, where it exists in lessened numbers, will, during fluctuations in the number of its enemies or of its prey, or in the seasons, be extremely liable to utter extermination; and thus its geographical range will come to be still more sharply defined.

If I am right in believing that allied or representative species, when inhabiting a continuous area, are generally so distributed that each has a wide range, with a comparatively narrow neutral territory between them, in which they become rather suddenly rarer and rarer; then, as varieties do not essentially differ from species, the same rule will probably apply to both; and if we in imagination adapt a varying species to a very large area, we shall have to adapt two varieties to two large areas, and a third variety to a narrow intermediate zone. The intermediate variety, consequently, will exist in lesser numbers from inhabiting a narrow and lesser area; and practically, as far as I can make out, this rule holds good with varieties in a state of nature. I have met with striking instances of the rule in the case of varieties intermediate between well-marked varieties in the genus Balanus. And it would appear from information given me by Mr. Watson, Dr. Asa Gray, and Mr. Wollaston, that generally when varieties intermediate between two other forms occur, they are much rarer numerically than the forms which they connect. Now, if we may trust these facts and inferences, and therefore conclude that varieties linking two other varieties together have generally existed in lesser numbers than the forms which they connect, then, I think, we can understand why intermediate varieties should not endure for very long periods;—why as a general rule they should be exterminated and disappear, sooner than the forms which they originally linked together.

For any form existing in lesser numbers would, as already remarked, run a greater chance of being exterminated than one existing in large numbers; and in this particular case the intermediate form would be eminently liable to the inroads of closely allied forms existing on both sides of it. But a far more important consideration, as I believe, is that, during the process of further modification, by which two varieties are supposed on my theory to be converted and perfected into two distinct species, the two which exist in larger numbers from inhabiting larger areas, will have a great advantage over the intermediate variety, which exists in smaller numbers in a narrow and intermediate zone. For forms existing in larger numbers will always have a better chance, within any given period, of presenting further favourable variations for natural selection to seize on, than will the rarer forms which exist in lesser numbers. Hence, the more common forms, in the race for life, will tend to beat and supplant the less common forms, for these will be more slowly modified and improved. It is the same principle which, as I believe, accounts for the common species in each country, as shown in the second chapter, presenting on an average a greater number of well-marked varieties than do the rarer species. I may illustrate what I mean by supposing three varieties of sheep to be kept, one adapted to an extensive mountainous region; a second to a comparatively narrow, hilly tract; and a third to wide plains at the base; and that the inhabitants are all trying with equal steadiness and skill to improve their stocks by selection; the chances in this case will be strongly in favour of the great holders on the mountains or on the plains improving their breeds more quickly than the small holders on the intermediate narrow, hilly tract; and consequently the improved mountain or plain breed will soon take the place of the less improved hill breed; and thus the two breeds, which originally existed in greater numbers, will come into close contact with each other, without the interposition of the supplanted, intermediate hill-variety.

To sum up, I believe that species come to be tolerably well-defined objects, and do not at any one period present an inextricable chaos of varying and intermediate links: firstly, because new varieties are very slowly formed, for variation is a very slow process, and natural selection can do nothing until favourable variations chance to occur, and until a place in the natural polity of the country can be better filled by some modification of some one or more of its inhabitants. And such new places will depend on slow changes of climate, or on the occasional immigration of new inhabitants, and, probably, in a still more important degree, on some of the old inhabitants becoming slowly modified, with the new forms thus produced and the old ones acting and reacting on each other. So that, in any one region and at any one time, we ought only to see a few species presenting slight modifications of structure in some degree permanent; and this assuredly we do see.

Secondly, areas now continuous must often have existed within the recent period in isolated portions, in which many forms, more especially amongst the classes which unite for each birth and wander much, may have separately been rendered sufficiently distinct to rank as representative species. In this case, intermediate varieties between the several representative species and their common parent, must formerly have existed in each broken portion of the land, but these links will have been supplanted and exterminated during the process of natural selection, so that they will no longer exist in a living state.

Thirdly, when two or more varieties have been formed in different portions of a strictly continuous area, intermediate varieties will, it is probable, at first have been formed in the intermediate zones, but they will generally have had a short duration. For these intermediate varieties will, from reasons already assigned (namely from what we know of the actual distribution of closely allied or representative species, and likewise of acknowledged varieties), exist in the intermediate zones in lesser numbers than the varieties which they tend to connect. From this cause alone the intermediate varieties will be liable to accidental extermination; and during the process of further modification through natural selection, they will almost certainly be beaten and supplanted by the forms which they connect; for these from existing in greater numbers will, in the aggregate, present more variation, and thus be further improved through natural selection and gain further advantages.

Lastly, looking not to any one time, but to all time, if my theory be true, numberless intermediate varieties, linking most closely all the species of the same group together, must assuredly have existed; but the very process of natural selection constantly tends, as has been so often remarked, to exterminate the parent-forms and the intermediate links. Consequently evidence of their former existence could be found only amongst fossil remains, which are preserved, as we shall in a future chapter attempt to show, in an extremely imperfect and intermittent record.

ON THE ORIGIN AND TRANSITIONS OF ORGANIC BEINGS WITH PECULIAR HABITS AND STRUCTURE

It has been asked by the opponents of such views as I hold, how, for instance, a land carnivorous animal could have been converted into one with aquatic habits; for how could the animal in its transitional state have subsisted? It would be easy to show that within the same group carnivorous animals exist having every intermediate grade between truly aquatic and strictly terrestrial habits; and as each exists by a struggle for life, it is clear that each is well adapted in its habits to its place in nature. Look at the Mustela vison of North America, which has webbed feet and which resembles an otter in its fur, short legs, and form of tail; during summer this animal dives for and preys on fish, but during the long winter it leaves the frozen waters, and preys like other pole-cats on mice and land animals. If a different case had been taken, and it had been asked how an insectivorous quadruped could possibly have been converted into a flying bat, the question would have been far more difficult, and I could have given no answer. Yet I think such difficulties have very little weight.

Here, as on other occasions, I lie under a heavy disadvantage, for out of the many striking cases which I have collected, I can give only one or two instances of transitional habits and structures in closely allied species of the same genus; and of diversified habits, either constant or occasional, in the same species. And it seems to me that nothing less than a long list of such cases is sufficient to lessen the difficulty in any particular case like that of the bat.

Look at the family of squirrels; here we have the finest gradation from animals with their tails only slightly flattened, and from others, as Sir J. Richardson has remarked, with the posterior part of their bodies rather wide and with the skin on their flanks rather full, to the so-called flying squirrels; and flying squirrels have their limbs and even the base of the tail united by a broad expanse of skin, which serves as a parachute and allows them to glide through the air to an astonishing distance from tree to tree. We cannot doubt that each structure is of use to each kind of squirrel in its own country, by enabling it to escape birds or beasts of prey, or to collect food more quickly, or, as there is reason to believe, by lessening the danger from occasional falls. But it does not follow from this fact that the structure of each squirrel is the best that it is possible to conceive under all natural conditions. Let the climate and vegetation change, let other competing rodents or new beasts of prey immigrate, or old ones become modified, and all analogy would lead us to believe that some at least of the squirrels would decrease in numbers or become exterminated, unless they also became modified and improved in structure in a corresponding manner. Therefore, I can see no difficulty, more especially under changing conditions of life, in the continued preservation of individuals with fuller and fuller flank-membranes, each modification being useful, each being propagated, until by the accumulated effects of this process of natural selection, a perfect so-called flying squirrel was produced.

Now look at the Galeopithecus or flying lemur, which formerly was falsely ranked amongst bats. It has an extremely wide flank-membrane, stretching from the corners of the jaw to the tail, and including the limbs and the elongated fingers: the flank-membrane is, also, furnished with an extensor muscle. Although no graduated links of structure, fitted for gliding through the air, now connect the Galeopithecus with the other Lemuridæ, yet I can see no difficulty in supposing that such links formerly existed, and that each had been formed by the same steps as in the case of the less perfectly gliding squirrels; and that each grade of structure had been useful to its possessor. Nor can I see any insuperable difficulty in further believing it possible that the membrane-connected fingers and fore-arm of the Galeopithecus might be greatly lengthened by natural selection; and this, as far as the organs of flight are concerned, would convert it into a bat. In bats which have the wing-membrane extended from the top of the shoulder to the tail, including the hind-legs, we perhaps see traces of an apparatus originally constructed for gliding through the air rather than for flight.

If about a dozen genera of birds had become extinct or were unknown, who would have ventured to have surmised that birds might have existed which used their wings solely as flappers, like the logger-headed duck (Micropterus of Eyton); as fins in the water and front legs on the land, like the penguin; as sails, like the ostrich; and functionally for no purpose, like the Apteryx. Yet the structure of each of these birds is good for it, under the conditions of life to which it is exposed, for each has to live by a struggle; but it is not necessarily the best possible under all possible conditions. It must not be inferred from these remarks that any of the grades of wing-structure here alluded to, which perhaps may all have resulted from disuse, indicate the natural steps by which birds have acquired their perfect power of flight; but they serve, at least, to show what diversified means of transition are possible.

Seeing that a few members of such water-breathing classes as the Crustacea and Mollusca are adapted to live on the land, and seeing that we have flying birds and mammals, flying insects of the most diversified types, and formerly had flying reptiles, it is conceivable that flying-fish, which now glide far through the air, slightly rising and turning by the aid of their fluttering fins, might have been modified into perfectly winged animals. If this had been effected, who would have ever imagined that in an early transitional state they had been inhabitants of the open ocean, and had used their incipient organs of flight exclusively, as far as we know, to escape being devoured by other fish?

When we see any structure highly perfected for any particular habit, as the wings of a bird for flight, we should bear in mind that animals displaying early transitional grades of the structure will seldom continue to exist to the present day, for they will have been supplanted by the very process of perfection through natural selection. Furthermore, we may conclude that transitional grades between structures fitted for very different habits of life will rarely have been developed at an early period in great numbers and under many subordinate forms. Thus, to return to our imaginary illustration of the flying-fish, it does not seem probable that fishes capable of true flight would have been developed under many subordinate forms, for taking prey of many kinds in many ways, on the land and in the water, until their organs of flight had come to a high stage of perfection, so as to have given them a decided advantage over other animals in the battle for life. Hence the chance of discovering species with transitional grades of structure in a fossil condition will always be less, from their having existed in lesser numbers, than in the case of species with fully developed structures.

I will now give two or three instances of diversified and of changed habits in the individuals of the same species. When either case occurs, it would be easy for natural selection to fit the animal, by some modification of its structure, for its changed habits, or exclusively for one of its several different habits. But it is difficult to tell, and immaterial for us, whether habits generally change first and structure afterwards; or whether slight modifications of structure lead to changed habits; both probably often change almost simultaneously. Of cases of changed habits it will suffice merely to allude to that of the many British insects which now feed on exotic plants, or exclusively on artificial substances. Of diversified habits innumerable instances could be given: I have often watched a tyrant flycatcher (Saurophagus sulphuratus) in South America, hovering over one spot and then proceeding to another, like a kestrel, and at other times standing stationary on the margin of water, and then dashing like a kingfisher at a fish. In our own country the larger titmouse (Parus major) may be seen climbing branches, almost like a creeper; it often, like a shrike, kills small birds by blows on the head; and I have many times seen and heard it hammering the seeds of the yew on a branch, and thus breaking them like a nuthatch. In North America the black bear was seen by Hearne swimming for hours with widely open mouth, thus catching, like a whale, insects in the water. Even in so extreme a case as this, if the supply of insects were constant, and if better adapted competitors did not already exist in the country, I can see no difficulty in a race of bears being rendered, by natural selection, more and more aquatic in their structure and habits, with larger and larger mouths, till a creature was produced as monstrous as a whale.

As we sometimes see individuals of a species following habits widely different from those both of their own species and of the other species of the same genus, we might expect, on my theory, that such individuals would occasionally have given rise to new species, having anomalous habits, and with their structure either slightly or considerably modified from that of their proper type. And such instances do occur in nature. Can a more striking instance of adaptation be given than that of a woodpecker for climbing trees and for seizing insects in the chinks of the bark? Yet in North America there are woodpeckers which feed largely on fruit, and others with elongated wings which chase insects on the wing; and on the plains of La Plata, where not a tree grows, there is a woodpecker, which in every essential part of its organisation, even in its colouring, in the harsh tone of its voice, and undulatory flight, told me plainly of its close blood-relationship to our common species; yet it is a woodpecker which never climbs a tree!

Petrels are the most aërial and oceanic of birds, yet in the quiet

Sounds of Tierra del Fuego, the Puffinuria berardi, in its general habits, in its astonishing power of diving, its manner of swimming, and of flying when unwillingly it takes flight, would be mistaken by any one for an auk or grebe; nevertheless, it is essentially a petrel, but with many parts of its organisation profoundly modified. On the other hand, the acutest observer by examining the dead body of the water-ouzel would never have suspected its sub-aquatic habits; yet this anomalous member of the strictly terrestrial thrush family wholly subsists by diving,—grasping the stones with its feet and using its wings under water.

He who believes that each being has been created as we now see it, must occasionally have felt surprise when he has met with an animal having habits and structure not at all in agreement. What can be plainer than that the webbed feet of ducks and geese are formed for swimming? yet there are upland geese with webbed feet which rarely or never go near the water; and no one except Audubon has seen the frigate-bird, which has all its four toes webbed, alight on the surface of the sea. On the other hand, grebes and coots are eminently aquatic, although their toes are only bordered by membrane. What seems plainer than that the long toes of grallatores are formed for walking over swamps and floating plants, yet the water-hen is nearly as aquatic as the coot; and the landrail nearly as terrestrial as the quail or partridge. In such cases, and many others could be given, habits have changed without a corresponding change of structure. The webbed feet of the upland goose may be said to have become rudimentary in function, though not in structure. In the frigate-bird, the deeply-scooped membrane between the toes shows that structure has begun to change.

He who believes in separate and innumerable acts of creation will say, that in these cases it has pleased the Creator to cause a being of one type to take the place of one of another type; but this seems to me only restating the fact in dignified language. He who believes in the struggle for existence and in the principle of natural selection, will acknowledge that every organic being is constantly endeavouring to increase in numbers; and that if any one being vary ever so little, either in habits or structure, and thus gain an advantage over some other inhabitant of the country, it will seize on the place of that inhabitant, however different it may be from its own place. Hence it will cause him no surprise that there should be geese and frigate-birds with webbed feet, either living on the dry land or most rarely alighting on the water; that there should be long-toed corncrakes living in meadows instead of in swamps; that there should be woodpeckers where not a tree grows; that there should be diving thrushes, and petrels with the habits of auks.

ORGANS OF EXTREME PERFECTION AND COMPLICATION

To suppose that the eye, with all its inimitable contrivances for adjusting the focus to different distances, for admitting different amounts of light, and for the correction of spherical and chromatic aberration, could have been formed by natural selection, seems, I freely confess, absurd in the highest possible degree. Yet reason tells me, that if numerous gradations from a perfect and complex eye to one very imperfect and simple, each grade being useful to its possessor, can be shown to exist; if further, the eye does vary ever so slightly, and the variations be inherited, which is certainly the case; and if any variation or modification in the organ be ever useful to an animal under changing conditions of life, then the difficulty of believing that a perfect and complex eye could be formed by natural selection, though insuperable by our imagination, can hardly be considered real. How a nerve comes to be sensitive to light, hardly concerns us more than how life itself first originated; but I may remark that several facts make me suspect that any sensitive nerve may be rendered sensitive to light, and likewise to those coarser vibrations of the air which produce sound.

In looking for the gradations by which an organ in any species has been perfected, we ought to look exclusively to its lineal ancestors; but this is scarcely ever possible, and we are forced in each case to look to species of the same group, that is to the collateral descendants from the same original parent-form, in order to see what gradations are possible, and for the chance of some gradations having been transmitted from the earlier stages of descent, in an unaltered or little altered condition. Amongst existing Vertebrata, we find but a small amount of gradation in the structure of the eye, and from fossil species we can learn nothing on this head. In this great class we should probably have to descend far beneath the lowest known fossiliferous stratum to discover the earlier stages, by which the eye has been perfected.

In the Articulata we can commence a series with an optic nerve merely coated with pigment, and without any other mechanism; and from this low stage, numerous gradations of structure, branching off in two fundamentally different lines, can be shown to exist, until we reach a moderately high stage of perfection. In certain crustaceans, for instance, there is a double cornea, the inner one divided into facets, within each of which there is a lens shaped swelling. In other crustaceans the transparent cones which are coated by pigment, and which properly act only by excluding lateral pencils of light, are convex at their upper ends and must act by convergence; and at their lower ends there seems to be an imperfect vitreous substance. With these facts, here far too briefly and imperfectly given, which show that there is much graduated diversity in the eyes of living crustaceans, and bearing in mind how small the number of living animals is in proportion to those which have become extinct, I can see no very great difficulty (not more than in the case of many other structures) in believing that natural selection has converted the simple apparatus of an optic nerve merely coated with pigment and invested by transparent membrane, into an optical instrument as perfect as is possessed by any member of the great Articulate class.

He who will go thus far, if he find on finishing this treatise that large bodies of facts, otherwise inexplicable, can be explained by the theory of descent, ought not to hesitate to go further, and to admit that a structure even as perfect as the eye of an eagle might be formed by natural selection, although in this case he does not know any of the transitional grades. His reason ought to conquer his imagination; though I have felt the difficulty far too keenly to be surprised at any degree of hesitation in extending the principle of natural selection to such startling lengths.

It is scarcely possible to avoid comparing the eye to a telescope. We know that this instrument has been perfected by the long-continued efforts of the highest human intellects; and we naturally infer that the eye has been formed by a somewhat analogous process. But may not this inference be presumptuous? Have we any right to assume that the Creator works by intellectual powers like those of man? If we must compare the eye to an optical instrument, we ought in imagination to take a thick layer of transparent tissue, with a nerve sensitive to light beneath, and then suppose every part of this layer to be continually changing slowly in density, so as to separate into layers of different densities and thicknesses, placed at different distances from each other, and with the surfaces of each layer slowly changing in form. Further we must suppose that there is a power always intently watching each slight accidental alteration in the transparent layers; and carefully selecting each alteration which, under varied circumstances, may in any way, or in any degree, tend to produce a distincter image. We must suppose each new state of the instrument to be multiplied by the million; and each to be preserved till a better be produced, and then the old ones to be destroyed. In living bodies, variation will cause the slight alterations, generation will multiply them almost infinitely, and natural selection will pick out with unerring skill each improvement. Let this process go on for millions on millions of years; and during each year on millions of individuals of many kinds; and may we not believe that a living optical instrument might thus be formed as superior to one of glass, as the works of the Creator are to those of man?

* * *

SUMMARY OF CHAPTER

We have in this chapter discussed some of the difficulties and objections which may be urged against my theory. Many of them are very grave; but I think that in the discussion light has been thrown on several facts, which on the theory of independent acts of creation are utterly obscure. We have seen that species at any one period are not indefinitely variable, and are not linked together by a multitude of intermediate gradations, partly because the process of natural selection will always be very slow, and will act, at any one time, only on a very few forms; and partly because the very process of natural selection almost implies the continual supplanting and extinction of preceding and intermediate gradations. Closely allied species, now living on a continuous area, must often have been formed when the area was not continuous, and when the conditions of life did not insensibly graduate away from one part to another. When two varieties are formed in two districts of a continuous area, an intermediate variety will often be formed, fitted for an intermediate zone; but from reasons assigned, the intermediate variety will usually exist in lesser numbers than the two forms which it connects; consequently the two latter, during the course of further modification, from existing in greater numbers, will have a great advantage over the less numerous intermediate variety, and will thus generally succeed in supplanting and exterminating it.

We have seen in this chapter how cautious we should be in concluding that the most different habits of life could not graduate into each other; that a bat, for instance, could not have been formed by natural selection from an animal which at first could only glide through the air.

We have seen that a species may under new conditions of life change its habits, or have diversified habits, with some habits very unlike those of its nearest congeners. Hence we can understand, bearing in mind that each organic being is trying to live wherever it can live, how it has arisen that there are upland geese with webbed feet, ground woodpeckers, diving thrushes, and petrels with the habits of auks.

Although the belief that an organ so perfect as the eye could have been formed by natural selection, is more than enough to stagger any one; yet in the case of any organ, if we know of a long series of gradations in complexity, each good for its possessor, then, under changing conditions of life, there is no logical impossibility in the acquirement of any conceivable degree of perfection through natural selection. In the cases in which we know of no intermediate or transitional states, we should be very cautious in concluding that none could have existed, for the homologies of many organs and their intermediate states show that wonderful metamorphoses in function are at least possible. For instance, a swim-bladder has apparently been converted into an air-breathing lung. The same organ having performed simultaneously very different functions, and then having been specialised for one function; and two very distinct organs having performed at the same time the same function, the one having been perfected whilst aided by the other, must often have largely facilitated transitions.

We are far too ignorant, in almost every case, to be enabled to assert that any part or organ is so unimportant for the welfare of a species, that modifications in its structure could not have been slowly accumulated by means of natural selection. But we may confidently believe that many modifications, wholly due to the laws of growth, and at first in no way advantageous to a species, have been subsequently taken advantage of by the still further modified descendants of this species. We may, also, believe that a part formerly of high importance has often been retained (as the tail of an aquatic animal by its terrestrial descendants), though it has become of such small importance that it could not, in its present state, have been acquired by natural selection,—a power which acts solely by the preservation of profitable variations in the struggle for life.

Natural selection will produce nothing in one species for the exclusive good or injury of another; though it may well produce parts, organs, and excretions highly useful or even indispensable, or highly injurious to another species, but in all cases at the same time useful to the owner. Natural selection in each well-stocked country, must act chiefly through the competition of the inhabitants one with another, and consequently will produce perfection, or strength in the battle for life, only according to the standard of that country. Hence the inhabitants of one country, generally the smaller one, will often yield, as we see they do yield, to the inhabitants of another and generally larger country. For in the larger country there will have existed more individuals, and more diversified forms, and the competition will have been severer, and thus the standard of perfection will have been rendered higher. Natural selection will not necessarily produce absolute perfection; nor, as far as we can judge by our limited faculties, can absolute perfection be everywhere found.

On the theory of natural selection we can clearly understand the full meaning of that old canon in natural history, "Natura non facit saltum." This canon, if we look only to the present inhabitants of the world, is not strictly correct, but if we include all those of past times, it must by my theory be strictly true.

It is generally acknowledged that all organic beings have been formed on two great laws—Unity of Type, and the Conditions of Existence. By unity of type is meant that fundamental agreement in structure, which we see in organic beings of the same class, and which is quite independent of their habits of life. On my theory, unity of type is explained by unity of descent. The expression of conditions of existence, so often insisted on by the illustrious Cuvier, is fully embraced by the principle of natural selection. For natural selection acts by either now adapting the varying parts of each being to its organic and inorganic conditions of life; or by having adapted them during long-past periods of time: the adaptations being aided in some cases by use and disuse, being slightly affected by the direct action of the external conditions of life, and being in all cases subjected to the several laws of growth. Hence, in fact, the law of the Conditions of Existence is the higher law; as it includes, through the inheritance of former adaptations, that of Unity of Type.

Chapter IX. On the Imperfection of the Geological Record

On the absence of intermediate varieties at the present day—On the nature of extinct intermediate varieties; on their number—On the vast lapse of time, as inferred from the rate of deposition and of denudation—On the poorness of our palæontological collections—On the intermittence of geological formations—On the absence of intermediate varieties in any one formation—On the sudden appearance of groups of species—On their sudden appearance in the lowest known fossiliferous strata.

In the sixth chapter I enumerated the chief objections which might be justly urged against the views maintained in this volume. Most of them have now been discussed. One, namely the distinctness of specific forms, and their not being blended together by innumerable transitional links, is a very obvious difficulty. I assigned reasons why such links do not commonly occur at the present day, under the circumstances apparently most favourable for their presence, namely on an extensive and continuous area with graduated physical conditions. I endeavoured to show, that the life of each species depends in a more important manner on the presence of other already defined organic forms, than on climate; and, therefore, that the really governing conditions of life do not graduate away quite insensibly like heat or moisture. I endeavoured, also, to show that intermediate varieties, from existing in lesser numbers than the forms which they connect, will generally be beaten out and exterminated during the course of further modification and improvement. The main cause, however, of innumerable intermediate links not now occurring everywhere throughout nature depends on the very process of natural selection, through which new varieties continually take the places of and exterminate their parent-forms. But just in proportion as this process of extermination has acted on an enormous scale, so must the number of intermediate varieties, which have formerly existed on the earth, be truly enormous. Why then is not every geological formation and every stratum full of such intermediate links? Geology assuredly does not reveal any such finely graduated organic chain; and this, perhaps, is the most obvious and gravest objection which can be urged against my theory. The explanation lies, as I believe, in the extreme imperfection of the geological record.

In the first place it should always be borne in mind what sort of intermediate forms must, on my theory, have formerly existed. I have found it difficult, when looking at any two species, to avoid picturing to myself, forms *directly* intermediate between them. But this is a wholly false view; we should always look for forms intermediate between each species and a common but unknown progenitor; and the progenitor will generally have differed in some respects from all its modified descendants. To give a simple illustration: the fantail and pouter pigeons have both descended from the rock-pigeon; if we possessed all the intermediate varieties which have ever existed, we should have an extremely close series between both and the rock-pigeon; but we should have no varieties directly intermediate between the fantail and pouter; none, for instance, combining a tail somewhat expanded with a crop somewhat enlarged, the characteristic features of these two breeds. These two breeds, moreover, have become so much modified, that if we had no historical or indirect evidence regarding their origin, it would not have been possible to have determined from a mere comparison of their structure with that of the rock-pigeon, whether they had descended from this species or from some other allied species, such as C. oenas.

So with natural species, if we look to forms very distinct, for instance to the horse and tapir, we have no reason to suppose that links ever existed directly intermediate between them, but between each and an unknown common parent. The common parent will have had in its whole organisation much general resemblance to the tapir and to the horse; but in some points of structure may have differed considerably from both, even perhaps more than they differ from each other. Hence in all such cases, we should be unable to recognise the parent-form of any two or more species, even if we closely compared the structure of the parent with that of its modified descendants, unless at the same time we had a nearly perfect chain of the intermediate links.

It is just possible by my theory, that one of two living forms might have descended from the other; for instance, a horse from a tapir; and in this case *direct* intermediate links will have existed between them. But such a case would imply that one form had remained for a very long period unaltered, whilst its descendants had undergone a vast amount of change; and the principle of competition between organism and organism, between child and parent, will render this a very rare event; for in all cases the new and improved forms of life will tend to supplant the old and unimproved forms.

By the theory of natural selection all living species have been connected with the parent-species of each genus, by differences not greater than we see between the varieties of the same species at the present day; and these parent-species, now generally extinct, have in their turn been similarly connected with more ancient species; and so on backwards, always converging to the common ancestor of each great class. So that the number of intermediate and transitional links, between all living and extinct species, must have been inconceivably great. But assuredly, if this theory be true, such have lived upon this earth.

ON THE LAPSE OF TIME

Independently of our not finding fossil remains of such infinitely numerous connecting links, it may be objected, that time will not have sufficed for so great an amount of organic change, all changes having been effected very slowly through natural selection. It is hardly possible for me even to recall to the reader, who may not be a practical geologist, the facts leading the mind feebly to comprehend the lapse of time. He who can read Sir Charles Lyell's grand work on the Principles of Geology, which the future historian will recognise as having produced a revolution in natural science, yet does not admit how incomprehensibly vast have been the past periods of time, may at once close this volume. Not that it suffices to study the Principles of Geology, or to read special treatises by different observers on separate formations, and to mark how each author attempts to give an inadequate idea of the duration of each formation or even each stratum. A man must for years examine for himself great piles of superimposed strata, and watch the sea at work grinding down old rocks and making fresh sediment, before he can hope to comprehend anything of the lapse of time, the monuments of which we see around us.

* * *

ON THE POORNESS OF OUR PALÆONTOLOGICAL COLLECTIONS

That our palæontological collections are very imperfect, is admitted by every one. The remark of that admirable palæontologist, the late Edward Forbes, should not be forgotten, namely, that numbers of our fossil species are known and named from single and often broken specimens, or from a few specimens collected on some one spot. Only a small portion of the surface of the earth has been geologically explored, and no part with sufficient care, as the important discoveries made every year in Europe prove. No organism wholly soft can be preserved. Shells and bones will decay and disappear when left on the bottom of the sea, where sediment is not accumulating. I believe we are continually taking a most erroneous view, when we tacitly admit to ourselves that sediment is being deposited over nearly the whole bed of the sea, at a rate sufficiently quick to embed and preserve fossil remains. Throughout an enormously large proportion of the ocean, the bright blue tint of the water bespeaks its purity. The many cases on record of a formation conformably covered, after an enormous interval of time, by another and later formation, without the underlying bed having suffered in the interval any wear and tear, seem explicable only on the view of the bottom of the sea not rarely lying for ages in an unaltered condition. The remains which do become embedded, if in sand or gravel, will when the beds are upraised generally be dissolved by the percolation of rain-water. I suspect that but few of the very many animals which live on the beach between high and low watermark are preserved. For instance, the several species of the Chthamalinæ (a sub-family of sessile cirripedes) coat the rocks all over the world in infinite numbers: they are all strictly littoral, with the exception of a single Mediterranean species, which inhabits deep water and has been found fossil in Sicily, whereas not one other species has hitherto been found in any tertiary formation: yet it is now known that the genus Chthamalus existed during the chalk period. The molluscan genus Chiton offers a partially analogous case.

With respect to the terrestrial productions which lived during the Secondary and Palæozoic periods, it is superfluous to state that our evidence from fossil remains is fragmentary in an extreme degree. For instance, not a land shell is known belonging to either of these vast periods, with one exception discovered by Sir C. Lyell in the carboniferous strata of North America. In regard to mammiferous remains, a single glance at the historical table published in the Supplement to Lyell's Manual, will bring home the truth, how accidental and rare is their preservation, far better than pages of detail. Nor is their rarity surprising, when we remember how large a proportion of the bones of tertiary mammals have been discovered either in caves or in lacustrine deposits; and that not a cave or true lacustrine bed is known belonging to the age of our secondary or palæozoic formations.

But the imperfection in the geological record mainly results from another and more important cause than any of the foregoing; namely, from the several formations being separated from each other by wide intervals of time. When we see the formations tabulated in written works, or when we follow them in nature, it is difficult to avoid believing that they are closely consecutive. But we know, for instance, from Sir R. Murchison's great work on Russia, what wide gaps there are in that country between the superimposed formations; so it is in North America, and in many other parts of the world. The most skilful geologist, if his attention had been exclusively confined to these large territories, would never have suspected that during the periods which were blank and barren in his own country, great piles of sediment, charged with new and peculiar forms of life, had elsewhere been accumulated. And if in each separate territory, hardly any idea can be formed of the length of time which has elapsed between the consecutive formations, we may infer that this could nowhere be ascertained. The frequent and great changes in the mineralogical composition of consecutive formations, generally implying great changes in the geography of the surrounding lands, whence the sediment has been derived, accords with the belief of vast intervals of time having elapsed between each formation.

* * *

Those who think the natural geological record in any degree perfect, and who do not attach much weight to the facts and arguments of other kinds given in this volume, will undoubtedly at once reject my theory. For my part, following out Lyell's metaphor, I look at the natural geological record, as a history of the world imperfectly kept, and written in a changing dialect; of this history we possess the last volume alone, relating only to two or three countries. Of this volume, only here and there a short chapter has been preserved;

and of each page, only here and there a few lines. Each word of the slowly-changing language, in which the history is supposed to be written, being more or less different in the interrupted succession of chapters, may represent the apparently abruptly changed forms of life, entombed in our consecutive, but widely separated, formations. On this view, the difficulties above discussed are greatly diminished, or even disappear.

Chapter XIII. Mutual Affinities of Organic Beings: Morphology; Embryology; Rudimentary Organs

CLASSIFICATION, groups subordinate to groups—Natural system—Rules and difficulties in classification, explained on the theory of descent with modification—Classification of varieties—Descent always used in classification—Analogical or adaptive characters—Affinities, general, complex and radiating—Extinction separates and defines groups—MORPHOLOGY, between members of the same class, between parts of the same individual—EMBRYOLOGY, laws of, explained by variations not supervening at an early age, and being inherited at a corresponding age—RUDIMENTARY ORGANS; their origin explained—Summary.

From the first dawn of life, all organic beings are found to resemble each other in descending degrees, so that they can be classed in groups under groups. This classification is evidently not arbitrary like the grouping of the stars in constellations. The existence of groups would have been of simple signification, if one group had been exclusively fitted to inhabit the land, and another the water; one to feed on flesh, another on vegetable matter, and so on; but the case is widely different in nature; for it is notorious how commonly members of even the same sub-group have different habits. In our second and fourth chapters, on Variation and on Natural Selection, I have attempted to show that it is the widely ranging, the much diffused and common, that is the dominant species belonging to the larger genera, which vary most. The varieties, or incipient species, thus produced ultimately become converted, as I believe, into new and distinct species; and these, on the principle of inheritance, tend to produce other new and dominant species. Consequently the groups which are now large, and which generally include many dominant species, tend to go on increasing indefinitely in size. I further attempted to show that from the varying descendants of each species trying to occupy as many and as different places as possible in the economy of nature, there is a constant tendency in their characters to diverge. This conclusion was supported by looking at the great diversity of the forms of life which, in any small area, come into the closest competition, and by looking to certain facts in naturalisation.

I attempted also to show that there is a constant tendency in the forms which are increasing in number and diverging in character, to supplant and exterminate the less divergent, the less improved, and preceding forms. I request the reader to turn to the diagram illustrating the action, as formerly explained, of these several principles; and he will see that the inevitable result is that the modified descendants proceeding from one progenitor become broken up into groups subordinate to groups. In the diagram each letter on the uppermost line may represent a genus including several species; and all the genera on this line form together one class, for all have descended from one ancient but unseen parent, and, consequently, have inherited something in common. But the three genera on the left hand have, on this same principle, much in common, and form a sub-family, distinct from that including the next two genera on the right hand, which diverged from a common parent at the fifth stage of descent. These five genera have also much, though less, in common; and they form a family distinct from that including the three genera still further to the right hand, which diverged at a still earlier period. And all these genera, descended from (A), form an order distinct from the genera descended from (I). So that we here have many species descended from a single progenitor grouped into genera; and the genera are included in, or subordinate to, sub-families, families, and orders, all united into one class. Thus, the grand fact in natural history of the subordination of group under group, which, from its familiarity, does not always sufficiently strike us, is in my judgment fully explained.

Naturalists try to arrange the species, genera, and families in each class, on what is called the Natural System. But what is meant by this system? Some authors look at it merely as a scheme for arranging together those living objects which are most alike, and for separating those which are most unlike; or as an artificial means for enunciating, as briefly as possible, general propositions,—that is, by one sentence to give the characters common, for instance, to all mammals, by another those common to all carnivora, by another those common to the dog-genus, and then by adding a single sentence, a full description is given of each kind of dog. The ingenuity and utility of this system are indisputable. But many naturalists think that something more is meant by the Natural System; they believe that it reveals the plan of the Creator; but unless it be specified whether order in time or space, or what else is meant by the plan of the Creator, it seems to me that nothing is thus added to our knowledge. Such expressions as that famous one of Linnæus, and which we often meet with in a more or less concealed form, that the characters do not make the genus, but that the genus gives the characters, seem to imply that something more is included in our classification, than mere resemblance. I believe that something more is included; and that propinquity of descent,—the only known cause of the similarity of organic beings,—is the bond, hidden as it is by various degrees of modification, which is partially revealed to us by our classifications.

* * *

But I must explain my meaning more fully. I believe that the *arrangement* of the groups within each class, in due subordination and relation to the other groups, must be strictly genealogical in order to be natural; but that the *amount* of difference in the several branches or groups, though allied in the same degree in blood to their common progenitor, may differ greatly, being due to the different degrees of modification which they have undergone; and this is expressed by the forms being ranked under different genera, families, sections, or orders. The reader will best understand what is meant, if he will take the trouble of referring to the diagram in the fourth chapter. We will suppose the letters A to L to represent allied genera, which lived during the Silurian epoch, and these have descended from a species which existed at an unknown anterior period. Species of three of these genera (A, F, and I) have transmitted modified descendants to the present day, represented by the fifteen genera (a^{14} to z^{14}) on the uppermost horizontal line. Now all these modified descendants from a single species, are represented as related in blood or descent to the same degree; they may metaphorically be called cousins to the same millionth degree; yet they differ widely and in different degrees from each other. The forms descended from A, now broken up into two or three families, constitute a distinct order from those descended from I, also broken up into two families. Nor can the existing species, descended from A, be ranked in the same genus with the parent A; or those from I, with the parent I. But the existing genus F^{14} may be supposed to have been but slightly modified; and it will then rank with the parent-genus F; just as some few still living organic beings belong to Silurian genera. So that the amount or value of the differences between organic beings all related to each other in the same degree in blood, has come to be widely different. Nevertheless their genealogical *arrangement* remains strictly true, not only at the present time, but at each successive period of descent. All the modified descendants from A will have inherited something in common from their common parent, as will all the descendants from I; so will it be with each subordinate branch of descendants, at each successive period. If, however, we choose to suppose that any of the descendants of A or of I have been so much modified as to have more or less completely lost traces of their parentage, in this case, their places in a natural classification will have been more or less completely lost,—as sometimes seems to have occurred with existing organisms. All the descendants of the genus F, along its whole line of descent, are supposed to have been but little modified, and they yet form a single genus. But this genus, though much isolated, will still occupy its proper intermediate position; for F originally was intermediate in character between A and I, and the several genera descended from these two genera will have inherited to a certain extent their characters. This natural arrangement is shown, as far as is possible on paper, in the diagram, but in much too simple a manner. If a branching diagram had not been used, and only the names of the groups had been written in a linear series, it would have been still less possible to have given a natural arrangement; and it is notoriously not possible to represent in a series, on a flat surface, the affinities which we discover in nature amongst the beings of the same group. Thus, on the view which I hold, the natural system is genealogical in its arrangement, like a pedigree; but the degrees of modification which the different groups have undergone, have to be expressed by ranking them under different so-called genera, sub-families, families, sections, orders, and classes.

* * *

On the principle of the multiplication and gradual divergence in character of the species descended from a common parent, together with their retention by inheritance of some characters in common, we can understand the excessively complex and radiating affinities by which all the members of the same family or higher group are connected together. For the common parent of a whole family of species, now broken up by extinction into distinct groups and sub-groups, will have transmitted some of its characters, modified in various ways and degrees, to all; and the several species will consequently be related to each other by circuitous lines of affinity of various lengths (as may be seen in the diagram so often referred to), mounting up through many predecessors. As it is difficult to show the blood-relationship between the numerous kindred of any ancient and noble family, even by the aid of a genealogical tree, and almost impossible to do this without this aid, we can understand the extraordinary difficulty which naturalists have experienced in describing, without the aid of a diagram, the various affinities which they perceive between the many living and extinct members of the same great natural class.

Extinction, as we have seen in the fourth chapter, has played an important part in defining and widening the intervals between the several groups in each class. We may thus account even for the distinctness of whole classes from each other—for instance, of birds from all other vertebrate animals—by the belief that many ancient forms of life have been utterly lost, through which the early progenitors of birds were formerly connected with the early progenitors of the other vertebrate classes. There has been less entire extinction of the forms of life which once connected fishes with batrachians. There has been still less in some other classes, as in that of the Crustacea, for here the most wonderfully diverse forms are still tied together by a long, but broken, chain of affinities. Extinction has only separated groups: it has by no means made them; for if every form which has ever lived on this earth were suddenly to reappear, though it would be quite impossible to give definitions by which each group could be distinguished from other groups, as all would blend together by steps as fine as those between the finest existing varieties, nevertheless a natural classification, or at least a natural arrangement, would be possible. * * *

Finally, we have seen that natural selection, which results from the struggle for existence, and which almost inevitably induces extinction and divergence of character in the many descendants from one dominant parent-species, explains that great and universal feature in the affinities of all organic beings, namely, their subordination in group under group. We use the element of descent in classing the individuals of both sexes and of all ages, although having few characters in common, under one species; we use descent in classing acknowledged varieties, however different they may be from their parent; and I believe this element of descent is the hidden bond of connexion which naturalists have sought under the term of the Natural System. On this idea of the natural system being, in so far as it has been perfected, genealogical in its arrangement, with the grades of difference between the descendants from a common parent, expressed by the terms genera, families, orders, &c., we can understand the rules which we are compelled to follow in our classification. We can understand why we value certain resemblances far more than others; why we are permitted to use rudimentary and useless organs, or others of trifling physiological importance; why, in comparing one group with a distinct group, we summarily reject analogical or adaptive characters, and yet use these same characters within the limits of the same group. We can clearly see how it is that all living and extinct forms can be grouped together in one great system; and how the several members of each class are connected together by the most complex and radiating lines of affinities. We shall never, probably, disentangle the inextricable web of affinities between the members of any one class; but when we have a distinct object in view, and do not look to some unknown plan of creation, we may hope to make sure but slow progress.

Morphology.—We have seen that the members of the same class, independently of their habits of life, resemble each other in the general plan of their organisation. This resemblance is often expressed by the term "unity of type;" or by saying that the several parts and organs in the different species of the class are homologous. The whole subject is included under the general name of Morphology. This is the most interesting department of natural history, and may be said to be its very soul. What can be more curious than that the hand of a man, formed for grasping, that of a mole for digging, the leg of the horse, the paddle of the porpoise, and the wing of the bat, should all be constructed on the same pattern, and should include the same bones, in the same relative positions? Geoffroy St. Hilaire has insisted strongly on the high importance of relative connexion in homologous organs: the parts may change to almost any extent in form and size, and yet they always remain connected together in the same order. We never find, for instance, the bones of the arm and forearm, or of the thigh and leg, transposed. Hence the same names can be given to the homologous bones in widely different animals. We see the same great law in the construction of the mouths of insects: what can be more different than the immensely long spiral proboscis of a sphinx-moth, the curious folded one of a bee or bug, and the great jaws of a beetle?—yet all these organs, serving for such different purposes, are formed by infinitely numerous modifications of an upper lip, mandibles, and two pairs of maxillæ. Analogous laws govern the construction of the mouths and limbs of crustaceans. So it is with the flowers of plants.

Nothing can be more hopeless than to attempt to explain this similarity of pattern in members of the same class, by utility or by the doctrine of final causes. The hopelessness of the attempt has been expressly admitted by Owen in his most interesting work on the 'Nature of Limbs.' On the ordinary view of the independent creation of each being, we can only say that so it is;—that it has so pleased the Creator to construct each animal and plant.

The explanation is manifest on the theory of the natural selection of successive slight modifications,—each modification being profitable in some way to the modified form, but often affecting by correlation of growth other parts of the organisation. In changes of this nature, there will be little or no tendency to modify the original pattern, or to transpose parts. The bones of a limb might be shortened and widened to any extent, and become gradually enveloped in thick membrane, so as to serve as a fin; or a webbed foot might have all its bones, or certain bones, lengthened to any extent, and the membrane connecting them increased to any extent, so as to serve as a wing: yet in all this great amount of modification there will be no tendency to alter the framework of bones or the relative connexion of the several parts. If we suppose that the ancient progenitor, the archetype as it may be called, of all mammals, had its limbs constructed on the existing general pattern, for whatever purpose they served, we can at once perceive the plain signification of the homologous construction of the limbs throughout the whole class. So with the mouths of insects, we have only to suppose that their common progenitor had an upper lip, mandibles, and two pair of maxillæ, these parts being perhaps very simple in form; and then natural selection will account for the infinite diversity in structure and function of the mouths of insects. Nevertheless, it is conceivable that the general pattern of an organ might become so much obscured as to be finally lost, by the atrophy and ultimately by the complete abortion of certain parts, by the soldering together of other parts, and by the doubling or multiplication of others,—variations which we know to be within the limits of possibility. In the paddles of the extinct gigantic sea-lizards, and in the mouths of certain suctorial crustaceans, the general pattern seems to have been thus to a certain extent obscured.

* * *

SUMMARY

In this chapter I have attempted to show, that the subordination of group to group in all organisms throughout all time; that the nature of the relationship, by which all living and extinct beings are united by complex, radiating, and circuitous lines of affinities into one grand system; the rules followed and the difficulties encountered by naturalists in their classifications; the value set upon characters, if constant and prevalent, whether of high vital importance, or of the most trifling importance, or, as in rudimentary organs, of no importance; the wide opposition in value between analogical or adaptive characters, and characters of true affinity; and other such rules;—all naturally follow on the view of the common parentage of those forms which are considered by naturalists as allied, together with their modification through natural selection, with its contingencies of extinction and divergence of character. In considering this view of classification, it should be borne in mind that the element of descent has been universally used in ranking together the sexes, ages, and acknowledged varieties of the same species, however different they may be in structure. If we extend the use of this element of descent,—the only certainly known cause of similarity in organic beings,—we shall understand what is meant by the natural system: it is genealogical in its attempted arrangement, with the grades of acquired difference marked by the terms varieties, species, genera, families, orders, and classes.

On this same view of descent with modification, all the great facts in Morphology become intelligible,—whether we look to the same pattern displayed in the homologous organs, to whatever purpose applied, of the different species of a class; or to the homologous parts constructed on the same pattern in each individual animal and plant.

On the principle of successive slight variations, not necessarily or generally supervening at a very early period of life, and being inherited at a corresponding period, we can understand the great leading facts in Embryology; namely, the resemblance in an individual embryo of the homologous parts, which when matured will become widely different from each other in structure and function; and the resemblance in different species of a class of the homologous parts or organs, though fitted in the adult members for purposes as different as possible. Larvae are active embryos, which have become specially modified in relation to their habits of life, through the principle of modifications being inherited at corresponding ages. On this same principle—and bearing in mind, that when organs are reduced in size, either from disuse or selection, it will generally be at that period of life when the being has to provide for its own wants, and bearing in mind how strong is the principle of inheritance—the occurrence of rudimentary organs and their final abortion, present to us no inexplicable difficulties; on the contrary, their presence might have been even anticipated. The importance of embryological characters and of rudimentary organs in classification is intelligible, on the view that an arrangement is only so far natural as it is genealogical.

Finally, the several classes of facts which have been considered in this chapter, seem to me to proclaim so plainly, that the innumerable species, genera, and families of organic beings, with which this world is peopled, have all descended, each within its own class or group, from common parents, and have all been modified in the course of descent, that I should without hesitation adopt this view, even if it were unsupported by other facts or arguments.

Chapter XIV. Recapitulation and Conclusion

Recapitulation of the difficulties on the theory of Natural Selection—Recapitulation of the general and special circumstances in its favour—Causes of the general belief in the immutability of species—How far the theory of natural selection may be extended—Effects of its adoption on the study of Natural history—Concluding remarks.

As this whole volume is one long argument, it may be convenient to the reader to have the leading facts and inferences briefly recapitulated.

That many and grave objections may be advanced against the theory of descent with modification through natural selection, I do not deny. I have endeavoured to give to them their full force. Nothing at first can appear more difficult to believe than that the more complex organs and instincts should have been perfected, not by means superior to, though analogous with, human reason, but by the accumulation of innumerable slight variations, each good for the individual possessor. Nevertheless, this difficulty, though appearing to our imagination insuperably great, cannot be considered real if we admit the following propositions, namely,—that gradations in the perfection of any organ or instinct, which we may consider, either do now exist or could have existed, each good of its kind,—that all organs and instincts are, in ever so slight a degree, variable,—and, lastly, that there is a struggle for existence leading to the preservation of each profitable deviation of structure or instinct. The truth of these propositions cannot, I think, be disputed.

It is, no doubt, extremely difficult even to conjecture by what gradations many structures have been perfected, more especially amongst broken and failing groups of organic beings; but we see so many strange gradations in nature, as is proclaimed by the canon, "Natura non facit saltum," that we ought to be extremely cautious in saying that any organ or instinct, or any whole being, could not have arrived at its present state by many graduated steps. There are, it must be admitted, cases of special difficulty on the theory of natural selection; and one of the most curious of these is the existence of two or three defined castes of workers or sterile females in the same community of ants; but I have attempted to show how this difficulty can be mastered.

With respect to the almost universal sterility of species when first crossed, which forms so remarkable a contrast with the almost universal fertility of varieties when crossed, I must refer the reader to the recapitulation of the facts given at the end of the eighth chapter, which seem to me conclusively to show that this sterility is no more a special endowment than is the incapacity of two trees to be grafted together; but that it is incidental on constitutional differences in the reproductive systems of the intercrossed species. We see the truth of this conclusion in the vast difference in the result, when the same two species are crossed reciprocally; that is, when one species is first used as the father and then as the mother.

The fertility of varieties when intercrossed and of their mongrel offspring cannot be considered as universal; nor is their very general fertility surprising when we remember that it is not likely that either their constitutions or their reproductive systems should have been profoundly modified. Moreover, most of the varieties which have been experimentised on have been produced under domestication; and as domestication apparently tends to eliminate sterility, we ought not to expect it also to produce sterility.

The sterility of hybrids is a very different case from that of first crosses, for their reproductive organs are more or less functionally impotent; whereas in first crosses the organs on both sides are in a perfect condition. As we continually see that organisms of all kinds are rendered in some degree sterile from their constitutions having been disturbed by slightly different and new conditions of life, we need not feel surprise at hybrids being in some degree sterile, for their constitutions can hardly fail to have been disturbed from being compounded of two distinct organisations. This parallelism is supported by another parallel, but directly opposite, class of facts; namely, that the vigour and fertility of all organic beings are increased by slight changes in their conditions of life, and that the offspring of slightly modified forms or varieties acquire from being crossed increased vigour and fertility. So that, on the one hand, considerable changes in the conditions of life and crosses between greatly modified forms, lessen fertility; and on the other hand, lesser changes in the conditions of life and crosses between less modified forms, increase fertility.

Turning to geographical distribution, the difficulties encountered on the theory of descent with modification are grave enough. All the individuals of the same species, and all the species of the same genus, or even higher group, must have descended from common parents; and therefore, in however distant and isolated parts of the world they are now found, they must in the course of successive generations have passed from some one part to the others. We are often wholly unable even to conjecture how this could have been effected. Yet, as we have reason to believe that some species have retained the same specific form for very long periods, enormously long as measured by years, too much stress ought not to be laid on the occasional wide diffusion of the same species; for during very long periods of time there will always be a good chance for wide migration by many means. A broken or interrupted range may often be accounted for by the extinction of the species in the intermediate regions. It cannot be denied that we are as yet very ignorant of the full extent of the various climatal and geographical changes which have affected the earth during modern periods; and such changes will obviously have greatly facilitated migration. As an example, I have attempted to show how potent has been the influence of the Glacial period on the distribution both of the same and of representative species throughout the world. We are as yet profoundly ignorant of the many occasional means of transport. With respect to distinct species of the same genus inhabiting very distant and isolated regions, as the process of modification has necessarily been slow, all the means of migration will have been possible during a very long period; and consequently the difficulty of the wide diffusion of species of the same genus is in some degree lessened.

As on the theory of natural selection an interminable number of intermediate forms must have existed, linking together all the species in each group by gradations as fine as our present varieties, it may be asked, Why do we not see these linking forms all around us? Why are not all organic beings blended together in an inextricable chaos? With respect to existing forms, we should remember that we have no right to expect (excepting in rare cases) to discover *directly* connecting links between them, but only between each and some extinct and supplanted form. Even on a wide area, which has during a long period remained continuous, and of which the climate and other conditions of life change insensibly in going from a district occupied by one species into another district occupied by a closely allied species, we have no just right to expect often to find intermediate varieties in the intermediate zone. For we have reason to believe that only a few species are undergoing change at any one period; and all changes are slowly effected. I have also shown that the intermediate varieties which will at first probably exist in the intermediate zones, will be liable to be supplanted by the allied forms on either hand; and the latter, from existing in greater numbers, will generally be modified and improved at a quicker rate than the intermediate varieties, which exist in lesser numbers; so that the intermediate varieties will, in the long run, be supplanted and exterminated.

On this doctrine of the extermination of an infinitude of connecting links, between the living and extinct inhabitants of the world, and at each successive period between the extinct and still older species, why is not every geological formation charged with such links? Why does not every collection of fossil remains afford plain evidence of the gradation and mutation of the forms of life? We meet with no such evidence, and this is the most obvious and forcible of the many objections which may be urged against my theory. Why, again, do whole groups of allied species appear, though certainly they often falsely appear, to have come in suddenly on the several geological stages? Why do we not find great piles of strata beneath the Silurian system, stored with the remains of the progenitors of the Silurian groups of fossils? For certainly on my theory such strata must somewhere have been deposited at these ancient and utterly unknown epochs in the world's history.

I can answer these questions and grave objections only on the supposition that the geological record is far more imperfect than most geologists believe. It cannot be objected that there has not been time sufficient for any amount of organic change; for the lapse of time has been so great as to be utterly inappreciable by the human intellect. The number of specimens in all our museums is absolutely as nothing compared with the countless generations of countless species which certainly have existed. We should not be able to recognise a species as the parent of any one or more species if we were to examine them ever so closely, unless we likewise possessed many of the intermediate links between their past or parent and present states; and these many links we could hardly ever expect to discover, owing to the imperfection of the geological record. Numerous existing doubtful forms could be named which are probably varieties; but who will pretend that in future ages so many fossil links will be discovered, that naturalists will be able to decide, on the common view, whether or not these doubtful forms are varieties? As long as most of the links between any two species are unknown, if any one link or intermediate variety be discovered, it will simply be classed as another and distinct species. Only a small portion of the world has been geologically explored. Only organic beings of certain classes can be preserved in a fossil condition, at least in any great number. Widely ranging species vary most, and varieties are often at first local,—both causes rendering the discovery of intermediate links less likely. Local varieties will not spread into other and distant regions until they are considerably modified and improved; and when they do spread, if discovered in a geological formation, they will appear as if suddenly created there, and will be simply classed as new species. Most formations have been intermittent in their accumulation; and their duration, I am inclined to believe, has been shorter than the average duration of specific forms. Successive formations are separated from each other by enormous blank intervals of time; for fossiliferous formations, thick enough to resist future degradation, can be accumulated only where much sediment is deposited on the subsiding bed of the sea. During the alternate periods of elevation and of stationary level the record will be blank. During these latter periods there will probably be more variability in the forms of life; during periods of subsidence, more extinction.

With respect to the absence of fossiliferous formations beneath the lowest Silurian strata, I can only recur to the hypothesis given in the ninth chapter. That the geological record is imperfect all will admit; but that it is imperfect to the degree which I require, few will be inclined to admit. If we look to long enough intervals of time, geology plainly declares that all species have changed; and they have changed in the manner which my theory requires, for they have changed slowly and in a graduated manner. We clearly see this in the fossil remains from consecutive formations invariably being much more closely related to each other, than are the fossils from formations distant from each other in time.

Such is the sum of the several chief objections and difficulties which may justly be urged against my theory; and I have now briefly recapitulated the answers and explanations which can be given to them. I have felt these difficulties far too heavily during many years to doubt their weight. But it deserves especial notice that the more important objections relate to questions on which we are confessedly ignorant; nor do we know how ignorant we are. We do not know all the possible transitional gradations between the simplest and the most perfect organs; it cannot be pretended that we know all the varied means of Distribution during the long lapse of years, or that we know how imperfect the Geological Record is. Grave as these several difficulties are, in my judgment they do not overthrow the theory of descent with modification.

Now let us turn to the other side of the argument. Under domestication we see much variability. This seems to be mainly due to the reproductive system being eminently susceptible to changes in the conditions of life; so that this system, when not rendered impotent, fails to reproduce offspring exactly like the parent-form. Variability is governed by many complex laws,—by correlation of growth, by use and disuse, and by the direct action of the physical conditions of life. There is much difficulty in ascertaining how much modification our domestic productions have undergone; but we may safely infer that the amount has been large, and that modifications can be inherited for long periods. As long as the conditions of life remain the same, we have reason to believe that a modification, which has already been inherited for many generations, may continue to be inherited for an almost infinite number of generations. On the other hand we have evidence that variability, when it has once come into play, does not wholly cease; for new varieties are still occasionally produced by our most anciently domesticated productions.

Man does not actually produce variability; he only unintentionally exposes organic beings to new conditions of life, and then nature acts on the organisation, and causes variability. But man can and does select the variations given to him by nature, and thus accumulate them in any desired manner. He thus adapts animals and plants for his own benefit or pleasure. He may do this methodically, or he may do it unconsciously by preserving the individuals most useful to him at the time, without any thought of altering the breed. It is certain that he can largely influence the character of a breed by selecting, in each successive generation, individual differences so slight as to be quite inappreciable by an uneducated eye. This process of selection has been the great agency in the production of the most distinct and useful domestic breeds. That many of the breeds produced by man have to a large extent the character of natural species, is shown by the inextricable doubts whether very many of them are varieties or aboriginal species.

There is no obvious reason why the principles which have acted so efficiently under domestication should not have acted under nature. In the preservation of favoured individuals and races, during the constantly-recurrent Struggle for Existence, we see the most powerful and ever-acting means of selection. The struggle for existence inevitably follows from the high geometrical ratio of increase which is common to all organic beings. This high rate of increase is proved by calculation, by the effects of a succession of peculiar seasons, and by the results of naturalisation, as explained in the third chapter. More individuals are born than can possibly survive. A grain in the balance will determine which individual shall live and which shall die,—which variety or species shall increase in number, and which shall decrease, or finally become extinct. As the individuals of the same species come in all respects into the closest competition with each other, the struggle will generally be most severe between them; it will be almost equally severe between the varieties of the same species, and next in severity between the species of the same genus. But the struggle will often be very severe between beings most remote in the scale of nature. The slightest advantage in one being, at any age or during any season, over those with which it comes into competition, or better adaptation in however slight a degree to the surrounding physical conditions, will turn the balance.

With animals having separated sexes there will in most cases be a struggle between the males for possession of the females. The most vigorous individuals, or those which have most successfully struggled with their conditions of life, will generally leave most progeny. But success will often depend on having special weapons or means of defence, or on the charms of the males; and the slightest advantage will lead to victory.

As geology plainly proclaims that each land has undergone great physical changes, we might have expected that organic beings would have varied under nature, in the same way as they generally have varied under the changed conditions of domestication. And if there be any variability under nature, it would be an unaccountable fact if natural selection had not come into play. It has often been asserted, but the assertion is quite incapable of proof, that the amount of variation under nature is a strictly limited quantity. Man, though acting on external characters alone and often capriciously, can produce within a short period a great result by adding up mere individual differences in his domestic productions; and every one admits that there are at least individual differences in species under nature. But, besides such differences, all naturalists have admitted the existence of varieties, which they think sufficiently distinct to be worthy of record in systematic works. No one can draw any clear distinction between individual differences and slight varieties; or between more plainly marked varieties and sub-species, and species. Let it be observed how naturalists differ in the rank which they assign to the many representative forms in Europe and North America.

If then we have under nature variability and a powerful agent always ready to act and select, why should we doubt that variations in any way useful to beings, under their excessively complex relations of life, would be preserved, accumulated, and inherited? Why, if man can by patience select variations most useful to himself, should nature fail in selecting variations useful, under changing conditions of life, to her living products? What limit can be put to this power, acting during long ages and rigidly scrutinising the whole constitution, structure, and habits of each creature,—favouring the good and rejecting the bad? I can see no limit to this power, in slowly and beautifully adapting each form to the most complex relations of life. The theory of natural selection, even if we looked no further than this, seems to me to be in itself probable. I have already recapitulated, as fairly as I could, the opposed difficulties and objections: now let us turn to the special facts and arguments in favour of the theory.

On the view that species are only strongly marked and permanent varieties, and that each species first existed as a variety, we can see why it is that no line of demarcation can be drawn between species, commonly supposed to have been produced by special acts of creation, and varieties which are acknowledged to have been produced by secondary laws. On this same view we can understand how it is that in each region where many species of a genus have been produced, and where they now flourish, these same species should present many varieties; for where the manufactory of species has been active, we might expect, as a general rule, to find it still in action; and this is the case if varieties be incipient species. Moreover, the species of the larger genera, which afford the greater number of varieties or incipient species, retain to a certain degree the character of varieties; for they differ from each other by a less amount of difference than do the species of smaller genera. The closely allied species also of the larger genera apparently have restricted ranges, and they are clustered in little groups round other species—in which respects they resemble varieties. These are strange relations on the view of each species having been independently created, but are intelligible if all species first existed as varieties.

As each species tends by its geometrical ratio of reproduction to increase inordinately in number; and as the modified descendants of each species will be enabled to increase by so much the more as they become more diversified in habits and structure, so as to be enabled to seize on many and widely different places in the economy of nature, there will be a constant tendency in natural selection to preserve the most divergent offspring of any one species. Hence during a long-continued course of modification, the slight differences, characteristic of varieties of the same species, tend to be augmented into the greater differences characteristic of species of the same genus. New and improved varieties will inevitably supplant and exterminate the older, less improved and intermediate varieties; and thus species are rendered to a large extent defined and distinct objects. Dominant species belonging to the larger groups tend to give birth to new and dominant forms; so that each large group tends to become still larger, and at the same time more divergent in character. But as all groups cannot thus succeed in increasing in size, for the world would not hold them, the more dominant groups beat the less dominant. This tendency in the large groups to go on increasing in size and diverging in character, together with the almost inevitable contingency of much extinction, explains the arrangement of all the forms of life, in groups subordinate to groups, all within a few great classes, which we now see everywhere around us, and which has prevailed throughout all time. This grand fact of the grouping of all organic beings seems to me utterly inexplicable on the theory of creation.

As natural selection acts solely by accumulating slight, successive, favourable variations, it can produce no great or sudden modification; it can act only by very short and slow steps. Hence the canon of "Natura non facit saltum," which every fresh addition to our knowledge tends to make more strictly correct, is on this theory simply intelligible. We can plainly see why nature is prodigal in variety, though niggard in innovation. But why this should be a law of nature if each species has been independently created, no man can explain.

Many other facts are, as it seems to me, explicable on this theory. How strange it is that a bird, under the form of woodpecker, should have been created to prey on insects on the ground; that upland geese, which never or rarely swim, should have been created with webbed feet; that a thrush should have been created to dive and feed on sub-aquatic insects; and that a petrel should have been created with habits and structure fitting it for the life of an auk or grebe! and so on in endless other cases. But on the view of each species constantly trying to increase in number, with natural selection always ready to adapt the slowly varying descendants of each to any unoccupied or ill-occupied place in nature, these facts cease to be strange, or perhaps might even have been anticipated.

As natural selection acts by competition, it adapts the inhabitants of each country only in relation to the degree of perfection of their associates; so that we need feel no surprise at the inhabitants of any one country, although on the ordinary view supposed to have been specially created and adapted for that country, being beaten and supplanted by the naturalised productions from another land. Nor ought we to marvel if all the contrivances in nature be not, as far as we can judge, absolutely perfect; and if some of them be abhorrent to our ideas of fitness. We need not marvel at the sting of the bee causing the bee's own death; at drones being produced in such vast numbers for one single act, and being then slaughtered by their sterile sisters; at the astonishing waste of pollen by our fir-trees; at the instinctive hatred of the queen bee for her own fertile daughters; at ichneumonidæ feeding within the live bodies of caterpillars; and at other such cases. The wonder indeed is, on the theory of natural selection, that more cases of the want of absolute perfection have not been observed.

The complex and little known laws governing variation are the same, as far as we can see, with the laws which have governed the production of so-called specific forms. In both cases physical conditions seem to have produced but little direct effect; yet when varieties enter any zone, they occasionally assume some of the characters of the species proper to that zone. In both varieties and species, use and disuse seem to have produced some effect; for it is difficult to resist this conclusion when we look, for instance, at the logger-headed duck, which has wings incapable of flight, in nearly the same condition as in the domestic duck; or when we look at the burrowing tucutucu, which is occasionally blind, and then at certain moles, which are habitually blind and have their eyes covered with skin; or when we look at the blind animals inhabiting the dark caves of America and Europe. In both varieties and species correlation of growth seems to have played a most important part, so that when one part has been modified other parts are necessarily modified. In both varieties and species reversions to long-lost characters occur. How inexplicable on the theory of creation is the occasional appearance of stripes on the shoulder and legs of the several species of the horse-genus and in their hybrids! How simply is this fact explained if we believe that these species have descended from a striped progenitor, in the same manner as the several domestic breeds of pigeon have descended from the blue and barred rock-pigeon!

On the ordinary view of each species having been independently created, why should the specific characters, or those by which the species of the same genus differ from each other, be more variable than the generic characters in which they all agree? Why, for instance, should the colour of a flower be more likely to vary in any one species of a genus, if the other species, supposed to have been created independently, have differently coloured flowers, than if all the species of the genus have the same coloured flowers? If species are only well-marked varieties, of which the characters have become in a high degree permanent, we can understand this fact; for they have already varied since they branched off from a common progenitor in certain characters, by which they have come to be specifically distinct from each other; and therefore these same characters would be more likely still to be variable than the generic characters which have been inherited without change for an enormous period. It is inexplicable on the theory of creation why a part developed in a very unusual manner in any one species of a genus, and therefore, as we may naturally infer, of great importance to the species, should be eminently liable to variation; but, on my view, this part has undergone, since the several species branched off from a common progenitor, an unusual amount of variability and modification, and therefore we might expect this part generally to be still variable. But a part may be developed in the most unusual manner, like the wing of a bat, and yet not be more variable than any other structure, if the part be common to many subordinate forms, that is, if it has been inherited for a very long period; for in this case it will have been rendered constant by long-continued natural selection.

Glancing at instincts, marvellous as some are, they offer no greater difficulty than does corporeal structure on the theory of the natural selection of successive, slight, but profitable modifications. We can thus understand why nature moves by graduated steps in endowing different animals of the same class with their several instincts. I have attempted to show how much light the principle of gradation throws on the admirable architectural powers of the hive-bee. Habit no doubt sometimes comes into play in modifying instincts; but it certainly is not indispensable, as we see, in the case of neuter insects, which leave no progeny to inherit the effects of long-continued habit. On the view of all the species of the same genus having descended from a common parent, and having inherited much in common, we can understand how it is that allied species, when placed under considerably different conditions of life, yet should follow nearly the same instincts; why the thrush of South America, for instance, lines her nest with mud like our British species. On the view of instincts having been slowly acquired through natural selection we need not marvel at some instincts being apparently not perfect and liable to mistakes, and at many instincts causing other animals to suffer.

If species be only well-marked and permanent varieties, we can at once see why their crossed offspring should follow the same complex laws in their degrees and kinds of resemblance to their parents,—in being absorbed into each other by successive crosses, and in other such points,—as do the crossed offspring of acknowledged varieties. On the other hand, these would be strange facts if species have been independently created, and varieties have been produced by secondary laws.

If we admit that the geological record is imperfect in an extreme degree, then such facts as the record gives, support the theory of descent with modification. New species have come on the stage slowly and at successive intervals; and the amount of change, after equal intervals of time, is widely different in different groups. The extinction of species and of whole groups of species, which has played so conspicuous a part in the history of the organic world, almost inevitably follows on the principle of natural selection; for old forms will be supplanted by new and improved forms. Neither single species nor groups of species reappear when the chain of ordinary generation has once been broken. The gradual diffusion of dominant forms, with the slow modification of their descendants, causes the forms of life, after long intervals of time, to appear as if they had changed simultaneously throughout the world. The fact of the fossil remains of each formation being in some degree intermediate in character between the fossils in the formations above and below, is simply explained by their intermediate position in the chain of descent. The grand fact that all extinct organic beings belong to the same system with recent beings, falling either into the same or into intermediate groups, follows from the living and the extinct being the offspring of common parents. As the groups which have descended from an ancient progenitor have generally diverged in character, the progenitor with its early descendants will often be intermediate in character in comparison with its later descendants; and thus we can see why the more ancient a fossil is, the oftener it stands in some degree intermediate between existing and allied groups. Recent forms are generally looked at as being, in some vague sense, higher than ancient and extinct forms; and they are in so far higher as the later and more improved forms have conquered the older and less improved organic beings in the struggle for life. Lastly, the law of the long endurance of allied forms on the same continent,—of marsupials in Australia, of edentata in America, and other such cases,—is intelligible, for within a confined country, the recent and the extinct will naturally be allied by descent.

Looking to geographical distribution, if we admit that there has been during the long course of ages much migration from one part of the world to another, owing to former climatal and geographical changes and to the many occasional and unknown means of dispersal, then we can understand, on the theory of descent with modification, most of the great leading facts in Distribution. We can see why there should be so striking a parallelism in the distribution of organic beings throughout space, and in their geological succession throughout time; for in both cases the beings have been connected by the bond of ordinary generation, and the means of modification have been the same. We see the full meaning of the wonderful fact, which must have struck every traveller, namely, that on the same continent, under the most diverse conditions, under heat and cold, on mountain and lowland, on deserts and marshes, most of the inhabitants within each great class are plainly related; for they will generally be descendants of the same progenitors and early colonists. On this same principle of former migration, combined in most cases with modification, we can understand, by the aid of the Glacial period, the identity of some few plants, and the close alliance of many others, on the most distant mountains, under the most different climates; and likewise the close alliance of some of the inhabitants of the sea in the northern and southern temperate zones, though separated by the whole intertropical ocean. Although two areas may present the same physical conditions of life, we need feel no surprise at their inhabitants being widely different, if they have been for a long period completely separated from each other; for as the relation of organism to organism is the most important of all relations, and as the two areas will have received colonists from some third source or from each other, at various periods and in different proportions, the course of modification in the two areas will inevitably be different.

On this view of migration, with subsequent modification, we can see why oceanic islands should be inhabited by few species, but of these, that many should be peculiar. We can clearly see why those animals which cannot cross wide spaces of ocean, as frogs and terrestrial mammals, should not inhabit oceanic islands; and why, on the other hand, new and peculiar species of bats, which can traverse the ocean, should so often be found on islands far distant from any continent. Such facts as the presence of peculiar species of bats, and the absence of all other mammals, on oceanic islands, are utterly inexplicable on the theory of independent acts of creation.

The existence of closely allied or representative species in any two areas, implies, on the theory of descent with modification, that the same parents formerly inhabited both areas; and we almost invariably find that wherever many closely allied species inhabit two areas, some identical species common to both still exist. Wherever many closely allied yet distinct species occur, many doubtful forms and varieties of the same species likewise occur. It is a rule of high generality that the inhabitants of each area are related to the inhabitants of the nearest source whence immigrants might have been derived. We see this in nearly all the plants and animals of the Galapagos archipelago, of Juan Fernandez, and of the other American islands being related in the most striking manner to the plants and animals of the neighbouring American mainland; and those of the Cape de Verde archipelago and other African islands to the African mainland. It must be admitted that these facts receive no explanation on the theory of creation.

The fact, as we have seen, that all past and present organic beings constitute one grand natural system, with group subordinate to group, and with extinct groups often falling in between recent groups, is intelligible on the theory of natural selection with its contingencies of extinction and divergence of character. On these same principles we see how it is, that the mutual affinities of the species and genera within each class are so complex and circuitous. We see why certain characters are far more serviceable than others for classification;—why adaptive characters, though of paramount importance to the being, are of hardly any importance in classification; why characters derived from rudimentary parts, though of no service to the being, are often of high classificatory value; and why embryological characters are the most valuable of all. The real affinities of all organic beings are due to inheritance or community of descent. The natural system is a genealogical arrangement, in which we have to discover the lines of descent by the most permanent characters, however slight their vital importance may be.

The framework of bones being the same in the hand of a man, wing of a bat, fin of the porpoise, and leg of the horse,—the same number of vertebræ forming the neck of the giraffe and of the elephant,—and innumerable other such facts, at once explain themselves on the theory of descent with slow and slight successive modifications. The similarity of pattern in the wing and leg of a bat, though used for such different purpose,—in the jaws and legs of a crab,—in the petals, stamens, and pistils of a flower, is likewise intelligible on the view of the gradual modification of parts or organs, which were alike in the early progenitor of each class. On the principle of successive variations not always supervening at an early age, and being inherited at a corresponding not early period of life, we can clearly see why the embryos of mammals, birds, reptiles, and fishes should be so closely alike, and should be so unlike the adult forms. We may cease marvelling at the embryo of an air-breathing mammal or bird having branchial slits and arteries running in loops, like those in a fish which has to breathe the air dissolved in water, by the aid of well-developed branchiæ.

Disuse, aided sometimes by natural selection, will often tend to reduce an organ, when it has become useless by changed habits or under changed conditions of life; and we can clearly understand on this view the meaning of rudimentary organs. But disuse and selection will generally act on each creature, when it has come to maturity and has to play its full part in the struggle for existence, and will thus have little power of acting on an organ during early life; hence the organ will not be much reduced or rendered rudimentary at this early age. The calf, for instance, has inherited teeth, which never cut through the gums of the upper jaw, from an early progenitor having well-developed teeth; and we may believe, that the teeth in the mature animal were reduced, during successive generations, by disuse or by the tongue and palate having been fitted by natural selection to browse without their aid; whereas in the calf, the teeth have been left untouched by selection or disuse, and on the principle of inheritance at corresponding ages have been inherited from a remote period to the present day. On the view of each organic being and each separate organ having been specially created, how utterly inexplicable it is that parts, like the teeth in the embryonic calf or like the shrivelled wings under the soldered wing-covers of some beetles, should thus so frequently bear the plain stamp of inutility! Nature may be said to have taken pains to reveal, by rudimentary organs and by homologous structures, her scheme of modification, which it seems that we wilfully will not understand.

I have now recapitulated the chief facts and considerations which have thoroughly convinced me that species have changed, and are still slowly changing by the preservation and accumulation of successive slight favourable variations. Why, it may be asked, have all the most eminent living naturalists and geologists rejected this view of the mutability of species? It cannot be asserted that organic beings in a state of nature are subject to no variation; it cannot be proved that the amount of variation in the course of long ages is a limited quantity; no clear distinction has been, or can be, drawn between species and well-marked varieties. It cannot be maintained that species when intercrossed are invariably sterile, and varieties invariably fertile; or that sterility is a special endowment and sign of creation. The belief that species were immutable productions was almost unavoidable as long as the history of the world was thought to be of short duration; and now that we have acquired some idea of the lapse of time, we are too apt to assume, without proof, that the geological record is so perfect that it would have afforded us plain evidence of the mutation of species, if they had undergone mutation.

But the chief cause of our natural unwillingness to admit that one species has given birth to other and distinct species, is that we are always slow in admitting any great change of which we do not see the intermediate steps. The difficulty is the same as that felt by so many geologists, when Lyell first insisted that long lines of inland cliffs had been formed, and great valleys excavated, by the slow action of the coast-waves. The mind cannot possibly grasp the full meaning of the term of a hundred million years; it cannot add up and perceive the full effects of many slight variations, accumulated during an almost infinite number of generations.

Although I am fully convinced of the truth of the views given in this volume under the form of an abstract, I by no means expect to convince experienced naturalists whose minds are stocked with a multitude of facts all viewed, during a long course of years, from a point of view directly opposite to mine. It is so easy to hide our ignorance under such expressions as the "plan of creation," "unity of design," &c., and to think that we give an explanation when we only restate a fact. Any one whose disposition leads him to attach more weight to unexplained difficulties than to the explanation of a certain number of facts will certainly reject my theory. A few naturalists, endowed with much flexibility of mind, and who have already begun to doubt on the immutability of species, may be influenced by this volume; but I look with confidence to the future, to young and rising naturalists, who will be able to view both sides of the question with impartiality. Whoever is led to believe that species are mutable will do good service by conscientiously expressing his conviction; for only thus can the load of prejudice by which this subject is overwhelmed be removed.

Several eminent naturalists have of late published their belief that a multitude of reputed species in each genus are not real species; but that other species are real, that is, have been independently created. This seems to me a strange conclusion to arrive at. They admit that a multitude of forms, which till lately they themselves thought were special creations, and which are still thus looked at by the majority of naturalists, and which consequently have every external characteristic feature of true species,—they admit that these have been produced by variation, but they refuse to extend the same view to other and very slightly different forms. Nevertheless they do not pretend that they can define, or even conjecture, which are the created forms of life, and which are those produced by secondary laws. They admit variation as a *vera causa* in one case, they arbitrarily reject it in another, without assigning any distinction in the two cases. The day will come when this will be given as a curious illustration of the blindness of preconceived opinion. These authors seem no more startled at a miraculous act of creation than at an ordinary birth. But do they really believe that at innumerable periods in the earth's history certain elemental atoms have been commanded suddenly to flash into living tissues? Do they believe that at each supposed act of creation one individual or many were produced? Were all the infinitely numerous kinds of animals and plants created as eggs or seed, or as full grown? and in the case of mammals, were they created bearing the false marks of nourishment from the mother's womb? Although naturalists very properly demand a full explanation of every difficulty from those who believe in the mutability of species, on their own side they ignore the whole subject of the first appearance of species in what they consider reverent silence.

It may be asked how far I extend the doctrine of the modification of species. The question is difficult to answer, because the more distinct the forms are which we may consider, by so much the arguments fall away in force. But some arguments of the greatest weight extend very far. All the members of whole classes can be connected together by chains of affinities, and all can be classified on the same principle, in groups subordinate to groups. Fossil remains sometimes tend to fill up very wide intervals between existing orders. Organs in a rudimentary condition plainly show that an early progenitor had the organ in a fully developed state; and this in some instances necessarily implies an enormous amount of modification in the descendants. Throughout whole classes various structures are formed on the same pattern, and at an embryonic age the species closely resemble each other. Therefore I cannot doubt that the theory of descent with modification embraces all the members of the same class. I believe that animals have descended from at most only four or five progenitors, and plants from an equal or lesser number.

Analogy would lead me one step further, namely, to the belief that all animals and plants have descended from some one prototype. But analogy may be a deceitful guide. Nevertheless all living things have much in common, in their chemical composition, their germinal vesicles, their cellular structure, and their laws of growth and reproduction. We see this even in so trifling a circumstance as that the same poison often similarly affects plants and animals; or that the poison secreted by the gall-fly produces monstrous growths on the wild rose or oak-tree. Therefore I should infer from analogy that probably all the organic beings which have ever lived on this earth have descended from some one primordial form, into which life was first breathed.

When the views entertained in this volume on the origin of species, or when analogous views are generally admitted, we can dimly foresee that there will be a considerable revolution in natural history. Systematists will be able to pursue their labours as at present; but they will not be incessantly haunted by the shadowy doubt whether this or that form be in essence a species. This I feel sure, and I speak after experience, will be no slight relief. The endless disputes whether or not some fifty species of British brambles are true species will cease. Systematists will have only to decide (not that this will be easy) whether any form be sufficiently constant and distinct from other forms, to be capable of definition; and if definable, whether the differences be sufficiently important to deserve a specific name. This latter point will become a far more essential consideration than it is at present; for differences, however slight, between any two forms, if not blended by intermediate gradations, are looked at by most naturalists as sufficient to raise both forms to the rank of species. Hereafter we shall be compelled to acknowledge that the only distinction between species and well-marked varieties is, that the latter are known, or believed, to be connected at the present day by intermediate gradations, whereas species were formerly thus connected. Hence, without quite rejecting the consideration of the present existence of intermediate gradations between any two forms, we shall be led to weigh more carefully and to value higher the actual amount of difference between them. It is quite possible that forms now generally acknowledged to be merely varieties may hereafter be thought worthy of specific names, as with the primrose and cowslip; and in this case scientific and common language will come into accordance. In short, we shall have to treat species in the same manner as those naturalists treat genera, who admit that genera are merely artificial combinations made for convenience. This may not be a cheering prospect; but we shall at least be freed from the vain search for the undiscovered and undiscoverable essence of the term species.

The other and more general departments of natural history will rise greatly in interest. The terms used by naturalists of affinity, relationship, community of type, paternity, morphology, adaptive characters, rudimentary and aborted organs, &c., will cease to be metaphorical, and will have a plain signification. When we no longer look at an organic being as a savage looks at a ship, as at something wholly beyond his comprehension; when we regard every production of nature as one which has had a history; when we contemplate every complex structure and instinct as the summing up of many contrivances, each useful to the possessor, nearly in the same way as when we look at any great mechanical invention as the summing up of the labour, the experience, the reason, and even the blunders of numerous workmen; when we thus view each organic being, how far more interesting, I speak from experience, will the study of natural history become!

A grand and almost untrodden field of inquiry will be opened, on the causes and laws of variation, on correlation of growth, on the effects of use and disuse, on the direct action of external conditions, and so forth. The study of domestic productions will rise immensely in value. A new variety raised by man will be a far more important and interesting subject for study than one more species added to the infinitude of already recorded species. Our classifications will come to be, as far as they can be so made, genealogies; and will then truly give what may be called the plan of creation. The rules for classifying will no doubt become simpler when we have a definite object in view. We possess no pedigrees or armorial bearings; and we have to discover and trace the many diverging lines of descent in our natural genealogies, by characters of any kind which have long been inherited. Rudimentary organs will speak infallibly with respect to the nature of long-lost structures. Species and groups of species, which are called aberrant, and which may fancifully be called living fossils, will aid us in forming a picture of the ancient forms of life. Embryology will reveal to us the structure, in some degree obscured, of the prototypes of each great class.

When we can feel assured that all the individuals of the same species, and all the closely allied species of most genera, have within a not very remote period descended from one parent, and have migrated from some one birthplace; and when we better know the many means of migration, then, by the light which geology now throws, and will continue to throw, on former changes of climate and of the level of the land, we shall surely be enabled to trace in an admirable manner the former migrations of the inhabitants of the whole world. Even at present, by comparing the differences of the inhabitants of the sea on the opposite sides of a continent, and the nature of the various inhabitants of that continent in relation to their apparent means of immigration, some light can be thrown on ancient geography.

The noble science of Geology loses glory from the extreme imperfection of the record. The crust of the earth with its embedded remains must not be looked at as a well-filled museum, but as a poor collection made at hazard and at rare intervals. The accumulation of each great fossiliferous formation will be recognised as having depended on an unusual concurrence of circumstances, and the blank intervals between the successive stages as having been of vast duration. But we shall be able to gauge with some security the duration of these intervals by a comparison of the preceding and succeeding organic forms. We must be cautious in attempting to correlate as strictly contemporaneous two formations, which include few identical species, by the general succession of their forms of life. As species are produced and exterminated by slowly acting and still existing causes, and not by miraculous acts of creation and by catastrophes; and as the most important of all causes of organic change is one which is almost independent of altered and perhaps suddenly altered physical conditions, namely, the mutual relation of organism to organism,—the improvement of one being entailing the improvement or the extermination of others; it follows, that the amount of organic change in the fossils of consecutive formations probably serves as a fair measure of the lapse of actual time. A number of species, however, keeping in a body might remain for a long period unchanged, whilst within this same period, several of these species, by migrating into new countries and coming into competition with foreign associates, might become modified; so that we must not overrate the accuracy of organic change as a measure of time. During early periods of the earth's history, when the forms of life were probably fewer and simpler, the rate of change was probably slower; and at the first dawn of life, when very few forms of the simplest structure existed, the rate of change may have been slow in an extreme degree. The whole history of the world, as at present known, although of a length quite incomprehensible by us, will hereafter be recognised as a mere fragment of time, compared with the ages which have elapsed since the first creature, the progenitor of innumerable extinct and living descendants, was created.

In the distant future I see open fields for far more important researches. Psychology will be based on a new foundation, that of the necessary acquirement of each mental power and capacity by gradation. Light will be thrown on the origin of man and his history.

Authors of the highest eminence seem to be fully satisfied with the view that each species has been independently created. To my mind it accords better with what we know of the laws impressed on matter by the Creator, that the production and extinction of the past and present inhabitants of the world should have been due to secondary causes, like those determining the birth and death of the individual. When I view all beings not as special creations, but as the lineal descendants of some few beings which lived long before the first bed of the Silurian system was deposited, they seem to me to become ennobled. Judging from the past, we may safely infer that not one living species will transmit its unaltered likeness to a distant futurity. And of the species now living very few will transmit progeny of any kind to a far distant futurity; for the manner in which all organic beings are grouped, shows that the greater number of species of each genus, and all the species of many genera, have left no descendants, but have become utterly extinct. We can so far take a prophetic glance into futurity as to foretel that it will be the common and widely-spread species, belonging to the larger and dominant groups, which will ultimately prevail and procreate new and dominant species. As all the living forms of life are the lineal descendants of those which lived long before the Silurian epoch, we may feel certain that the ordinary succession by generation has never once been broken, and that no cataclysm has desolated the whole world. Hence we may look with some confidence to a secure future of equally inappreciable length. And as natural selection works solely by and for the good of each being, all corporeal and mental endowments will tend to progress towards perfection.

It is interesting to contemplate an entangled bank, clothed with many plants of many kinds, with birds singing on the bushes, with various insects flitting about, and with worms crawling through the damp earth, and to reflect that these elaborately constructed forms, so different from each other, and dependent on each other in so complex a manner, have all been produced by laws acting around us. These laws, taken in the largest sense, being Growth with Reproduction; Inheritance which is almost implied by reproduction; Variability from the indirect and direct action of the external conditions of life, and from use and disuse; a Ratio of Increase so high as to lead to a Struggle for Life, and as a consequence to Natural Selection, entailing Divergence of Character and the Extinction of less-improved forms. Thus, from the war of nature, from famine and death, the most exalted object which we are capable of conceiving, namely, the production of the higher animals, directly follows. There is grandeur in this view of life, with its several powers, having been originally breathed into a few forms or into one; and that, whilst this planet has gone cycling on according to the fixed law of gravity, from so simple a beginning endless forms most beautiful and most wonderful have been, and are being, evolved.

Selected Readings

DARWIN'S LIFE

Barlow, Nora, ed. *The Autobiography of Charles Darwin, 1809–1882,* (with original omissions restored). New York, 1969.
Bowlby, J. *Charles Darwin: A Biography.* London, 1990.
Brent, Peter. *Charles Darwin.* London, 1981.
Browne, Janet. *Charles Darwin, I: Voyaging.* New York, 1995.
Burckhardt, Frederick, Sydney Smith, and others, eds. *The Correspondence of Charles Darwin,* vols. 1–10. Cambridge, UK, 1985–.
Colp, Ralph, Jr. *To Be an Invalid: The Illness of Charles Darwin.* Chicago, 1977.
Darwin, Francis, ed. *The Life and Letters of Charles Darwin.* New York, 1959.
DeBeer, Gavin. *Charles Darwin: A Scientific Biography.* New York, 1964.
Desmond, Adrian, and James Moore. *Darwin.* London, 1991.
Goldstein, Jared Haft. "Darwin, Chagas', Mind, and Body." *Perspectives in Biology and Medicine* 32 (1989): 586–601.
Gruber, Howard E. *Darwin on Man: A Psychological Study of Scientific Creativity.* Chicago, 1981.
Litchfield, H. E. *Emma Darwin: A Century of Family Letters, 1792–1896.* London, 1915.
Moore, James. *The Darwin Legend.* Grand Rapids, Mich., 1994.

SCIENTIFIC THOUGHT: JUST BEFORE DARWIN

Chambers, Robert. *Vestiges of the Natural History of Creation.* London, 1844.
Herschel, John. *Preliminary Discourse on the Study of Natural Philosophy.* London, 1830.
Lamarck, Jean Baptiste Pierre Antoine de Monet. *Zoological Philosophy.* Paris, 1809.
Lyell, Charles. *Principles of Geology.* London, 1830–33.
Malthus, Thomas Robert. *An Essay on the Principle of Population.* London, 1798; rev., 1803.
Mill, John Stuart. *A System of Logic, Ratiocination and Induction.* London, 1843.
Paley, William. *Natural Theology: Or Evidences of the Existence and Attributes of the Deity Collected from the Appearances of Nature.* London, 1802.
Wallace, Alfred Russel. "On the Tendency of Varieties to Depart Indefinitely from the Original Type." *Journal of the Proceedings of the Linnean Society, Zoology* III (August 20, 1858).
Whewell, William. *Astronomy and General Physics Considered with Reference to Natural Theology.* London, 1833.

SELECTIONS FROM DARWIN'S WORK

Barrett, Paul, and others, eds. *Charles Darwin's Notebooks, 1836–1844.* Cambridge, UK, 1985.
Burkhardt, Frederick, and others. *Darwin's Scientific Diaries, 1836–1842.* Cambridge, UK, 1987.
Darwin, Charles. *Journal of Researches into the Geology and Natural History of the Various Countries Visited by H.M.S. "Beagle"* . . . London, 1839; 2nd ed., 1845.
———. *The Structure and Distribution of Coral Reefs.* London, 1842.
———. *A Monograph of the Sub-class Cirripedia* . . . London, 1851.
———. "On the Tendency of Species to Form Varieties . . ." *Journal of the Proceedings of the Linnean Society, Zoology* III (August 20, 1858): 45–62.
———. *On the Origin of Species by Means of Natural Selection* . . . London, 1859; facsimile edition, Cambridge, Mass., 1964.

———. *On the Various Contrivances by Which British and Foreign Orchids Are Fertilized by Insects, and on the Good Effects of Intercrossing.* London, 1862.
———. *The Variation of Animals and Plants under Domestication.* London, 1868.
———. *The Descent of Man* . . . London, 1871.
———. *The Expression of the Emotions in Man and Animals.* London, 1872.
———. *Insectivorous Plants.* London, 1875.
———. *The Power of Movements in Plants.* London, 1880.
———. *The Formation of Vegetable Mould* . . . London, 1881.
Di Gregorio, Mario A., ed. *Charles Darwin's Marginalia, I.* 1990.
Keynes, Richard, ed. *Charles Darwin's "Beagle" Diary.* Cambridge, UK, 1988.
Stauffer, R. C., ed. *Charles Darwin's Natural Selection: Being the Second Part of His Big Species Book Written from 1856 to 1858.* Cambridge, UK, 1975.

DARWIN'S INFLUENCE ON SCIENCE

Ayala, Francisco J., ed. *Molecular Evolution.* New York, 1976.
Barkow, Jerome H., Leda Cosmides, and John Tooby. *The Adapted Mind: Evolutionary Psychology and the Generation of Culture.* New York, 1992.
Blinderman, Charles. *The Piltdown Inquest.* Buffalo, N.Y., 1986.
Bowler, Peter J. *Darwinism.* New York, 1993.
———. *Evolution: The History of an Idea.* Berkeley, Calif., 1984.
Clark, W. E. LeGros. *The Fossil Evidence for Human Evolution.* Chicago, 1978.
Corballis, Michael C. *The Lopsided Ape: Evolution of the Generative Mind.* New York, 1991.
Dawkins, Richard. *The Selfish Gene.* New York, 1976.
———. *The Blind Watchmaker.* London, 1986.
———. *Climbing Mount Improbable.* New York, 1996.
———. *Unweaving the Rainbow.* New York, 1998.
Depew, David J., and Bruce H. Weber. *Darwinism Evolving: Systems Dynamics and the Genealogy of Natural Selection.* Cambridge, Mass., 1995.
Desmond, Adrian. *Huxley.* Reading, Mass., 1997.
Dobzhansky, Theodosius. *Genetics and the Origin of Species.* New York, 1937.
Durant, John R., ed. *Human Origins.* New York, 1989.
Eibl-Eibesfeldt, Irenaus. *Human Ethology.* Hawthorne, N.Y., 1989.
Eiseley, Loren. *Darwin's Century.* New York, 1958.
Eldredge, Niles. *Reinventing Darwin* . . . London, 1995.
Endler, John A. *Natural Selection in the Wild.* Princeton, N.J., 1986.
Fisher, Ronald A. *The Genetical Theory of Natural Selection.* Oxford, UK, 1930.
Flanagan, Dennis, and others. *Evolution.* New York, 1978.
Fossey, Dian. *Gorillas in the Mist.* Boston, 1983.
Futuyma, D. J. *Evolutionary Biology.* Sunderland, Mass., 1979.
Ghiselin, Michael T. *The Triumph of the Darwinian Method.* Berkeley, Calif., 1969.
Gillespie, C. C. *Genesis and Geology.* Cambridge, Mass., 1959.
Goodall, Jane. *The Chimpanzees of Gombe.* Cambridge, Mass., 1986.
———. *In the Shadow of Man.* Boston, 1983.
Gould, James L., and William T. Keeton, with Carol Grant Gould. *Biological Science.* New York, 1996.
Gould, Stephen Jay. *Ever Since Darwin.* New York, 1977.
———. *Ontogeny and Phylogeny.* Cambridge, Mass., 1977.
———. *The Panda's Thumb.* New York, 1980.
———. *An Urchin in the Storm.* New York, 1987.
———. *Full House.* New York, 1996.
——— and Niles Eldredge. "Punctuated Equilibrium: An Alternative to Phyletic Gradualism." In T. J. M. Schopf, ed. *Models in Paleobiology.* San Francisco, 1972.
Grant, Peter R. *Ecology and Evolution of Darwin's Finches.* Princeton, N.J., 1986.
——— and Rosemary Grant. *Evolutionary Dynamics of a Natural Population: The Large Cactus Finch of the Galapagos.* Chicago, 1989.
Gruber, Howard. *Darwin on Man.* London, 1974.
Haraway, Donna. *Primate Visions.* New York, 1989.
Hennig, Willi. *Phylogenetic Systematics.* Trans. D. Dwight Davis and Rainer Zangerl. Urbana, Ill., 1966.
Hull, David L. *Darwin and His Critics.* Chicago, 1973.
Huxley, Julian. *Evolution: The Modern Synthesis.* New York, 1963.
———. *Evolution in Action.* New York, 1953.
Jann, Rosemary. "Darwin and the Anthropologists: Sexual Selection and Its Discontents." *Victorian Studies* 37 (1984): 286–306.

Johanson, Donald C., and Maitland A. Edey. *Lucy: The Beginnings of Humankind.* New York, 1982.
——. *Lucy's Child: The Discovery of a Human Ancestor.* New York, 1989.
——. *Blueprints: Solving the Mystery of Evolution.* Boston, 1989.
—— and Greta Jones. *Social Darwinism and English Thought.* Brighton, UK, 1980.
Koertge, Noretta. *A House Built on Sand.* New York, 1998.
Kuper, Adam. *The Chosen Primate,* Cambridge, Mass., 1994.
Kohn, David, ed. *The Darwinian Heritage.* Princeton, N.J., 1985.
Lancaster, Jane B. *Primate Behavior and the Emergence of Human Culture.* New York, 1975.
Leakey, Louis S. B. *The Progress and Evolution of Man in Africa.* New York, 1961.
Leakey, Mary D. *Disclosing the Past: An Autobiography.* New York, 1985.
Leakey, Richard. *Origins Reconsidered: In Search of What Makes Us Human.* New York, 1992.
—— and Roger Lewin. *The Sixth Extinction: Patterns of Life and the Future of Humankind.* New York, 1995.
Lewin, Roger. *Thread of Life: The Smithsonian Looks at Evolution.* Washington, D.C., 1982.
Margulis, Lynn. *Symbiosis in Cell Evolution.* New York, 1993.
Matsumura, Molleen, ed. *Voices for Evolution.* 2nd ed. Berkeley, Calif., 1995.
Mayr, Ernst. *The Growth of Biological Thought.* Cambridge, Mass., 1982.
——. *One Long Argument: Charles Darwin and the Genesis of Evolutionary Thought.* Cambridge, Mass., 1991.
——. *This Is Biology: The Science of the Living World.* Cambridge, Mass., 1997.
Mead, Margaret. *Continuities in Cultural Evolution.* New Haven, Conn., 1964.
Montgomery, Sy. *Walking with the Great Apes: Jane Goodall, Dian Fossey, Birute Galdikas.* Boston, 1992.
Morgan, T. H. *Evolution and Genetics.* Princeton, N.J., 1925.
National Academy of Sciences. *Teaching about Evolution and the Nature of Science.* Washington, D.C., 1998.
——. *Science and Creationism.* Washington, D.C., 1999.
Oldroyd, David R. *Darwinian Impacts: An Introduction to the Darwinian Revolution.* Atlantic Highlands, N.J., 1980.
Ospovat, Dov. *The Development of Darwin's Theory: Natural History, Natural Theology, and Natural Selection.* Cambridge, UK, 1981.
Otte, Daniel, and John A. Endler, eds. *Speciation and Its Consequences.* Sunderland, Mass., 1989.
Pinker, Steven, *The Language Instinct: How the Mind Creates Language.* New York, 1994.
——. *How the Mind Works.* New York, 1997.
Richards, Robert J. *The Meaning of Evolution.* Chicago, 1992.
Ruse, Michael. *The Darwinian Revolution.* Chicago, 1979.
——. *The Darwinian Paradigm: Essays on Its History, Philosophy, and Religious Implications.* London, 1989.
——. *Monad to Man: The Concept of Progress in Evolutionary Biology.* Cambridge, Mass., 1996.
Smith, Fred H., and Frank Spencer, eds. *The Origins of Modern Humans: A World Survey of the Fossil Evidence.* New York, 1984.
Smith, John Maynard. *The Theory of Evolution.* Harmondsworth, UK, 1975.
Simpson, George Gaylord. *The Meaning of Evolution.* New Haven, Conn., 1949.
Sober, Elliott. *Conceptual Issues in Evolutionary Biology.* Cambridge, Mass., 1984.
——. *The Nature of Selection.* Cambridge, Mass., 1985.
Smuts, Barbara B. *Sex and Friendship in Baboons.* New York, 1985.
Stringer, Christopher, and Robin McKie. *African Exodus: The Origins of Modern Humanity.* New York, 1997.
Tattersall, Ian. *Becoming Human.* New York, 1998.
——. *The Human Trail: How We Know What We Think We Know about Human Evolution.* New York, 1995.
——, Eric Delson, and John Van Couvering. *Encyclopedia of Human Evolution and Prehistory.* New York, 1988.
Tax, Sol. *Evolution after Darwin.* Chicago, 1960.
Weiner, Jonathan. *The Beak of the Finch.* New York, 1994.
Wilson, Edward O. *Sociobiology.* Cambridge, Mass., 1975.
——. *Consilience.* New York, 1998.
Williams, George C. *Natural Selection: Domains, Levels, and Applications.* New York, 1992.

DARWINIAN PATTERNS IN SOCIAL THOUGHT

Barlow, Connie. *Green Space, Green Time.* New York, 1997.
Birken, Lawrence. "Darwin and Gender." *Proteus* 6 (1989): 24–29.
Bowler, Peter J. *Biology and Social Thought, 1850–1914.* Berkeley, Calif., 1993.
Churchill, Frederick B. "Darwin and the Historians." In R. J. Berry, ed. *Charles Darwin: A Commemoration, 1882–1982.* London, 1982.
Degler, Carl N. *In Search of Human Nature: The Decline and Revival of Darwinism in American Social Thought.* New York, 1991.
Carnegie, Andrew. *The Gospel of Wealth . . .* New York, 1900.
Diamond, Jared. *The Third Chimpanzee: The Evolution and Future of the Human Animal.* New York, 1992.
Ehrenreich, Barbara. *Blood Rites: Origins and History of the Passions of War.* New York, 1997.
—— and Janet McIntosh. "The New Creationism: Biology under Attack." *The Nation* (June 9, 1997): 11–16.
Gordon, Scott. "Darwin and Political Economy: The Connection Reconsidered." *Journal of the History of Biology* 22 (1989): 437–59.
Gould, Stephen Jay. *The Mismeasure of Man.* New York, 1977.
Greene, John C. *The Death of Adam: Evolution and Its Impact on Western Thought.* New York, 1959.
Hrdy, Sarah Blaffer. *Mother Nature: A History of Mothers, Infants, and Natural Selection.* New York, 1999.
Hofstadter, Richard. *Social Darwinism in American Thought.* Boston, 1955.
Kropotkin Peter. *Mutual Aid.* London, 1902.
Lewontin, R. C., Steven Rose, and Leon J. Kamin. *Not in Our Genes.* New York, 1984.
Langs, Robert. *The Evolution of the Emotion-Processing Mind.* London, 1996.
Mayr, Ernst. *Evolution and the Diversity of Life.* Cambridge, Mass., 1976.
Nesse, Randolph, and George Williams. *Why We Get Sick: The New Science of Darwinian Medicine.* New York, 1994.
Oldroyd, David, and Ian Langham. *The Wider Domain of Evolutionary Thought.* London, 1983.
Stanton, Elizabeth Cady. *The Woman's Bible* (1895, 1898). See Annie Laurie Gaylor, ed. *Women without Superstition.* Madison, Wisc., 1997.
Tanner, Nancy M. *On Becoming Human.* Cambridge, UK, 1981.

DARWINIAN INFLUENCES IN PHILOSOPHY AND ETHICS

Birx, H. James. *Theories of Evolution.* Springfield, Ill., 1984.
Dennett, Daniel. *Darwin's Dangerous Idea: Evolution and the Meanings of Life.* London, 1995.
DeWaal, Frans. *Good Natured.* Cambridge, Mass., 1996.
Dewey, John. *The Influence of Darwin on Philosophy . . .* New York, 1910.
Gillispie, C. C. *Charles Darwin and the Problem of Creation.* Chicago, 1979.
Gorney, Roderic. *The Human Agenda.* Los Angeles, 1979.
Hull, David L. *Philosophy of Biological Science.* Englewood Cliffs, N.J., 1974.
Huxley, Thomas Henry, and Julian Huxley. *Touchstone for Ethics.* New York, 1947.
Mayr, Ernst. "Darwin's Impact on Modern Thought." *Proceedings of the American Philosophical Society* 139 (1995): 317–25.
Rachels, James. *Created from Animals: The Moral Implications of Darwinism.* New York, 1990.
Ridley, Matt. *The Origins of Virtue: Human Instincts and the Evolution of Cooperation.* New York, 1997.
Ruse, Michael. *Philosophy of Biology.* Buffalo, N.Y., 1998.
——. *Taking Darwin Seriously: A Naturalistic Approach to Philosophy.* Oxford, UK, 1987.
Waddington, C. H. *The Ethical Animal.* London, 1960.
Wilson, James Q. *The Moral Sense.* New York, 1993.

EVOLUTIONARY THEORY AND RELIGIOUS THEORY

Behe, Michael J. *Darwin's Black Box: The Biochemical Challenge to Evolution.* New York, 1996.
Birx, H. James. "Origin of Life and Unbelief." In Gordon Stein, ed. *Encyclopedia of Unbelief.* Buffalo, N.Y., 1985.

Edis, Taner. "Islamic Creationism in Turkey." *Creation/Evolution* 34 (1994): 1–4.
Eldredge, Niles. *The Monkey Business: A Scientist Looks at Creationism.* New York, 1982.
Eve, Raymond A., and Francis B. Harrold. *The Creationist Movement in Modern America.* Boston, 1991.
———, eds. *Cult Archaeology and Creationism: Understanding Pseudoscientific Beliefs about the Past.* Iowa City, 1987.
Frye, R. *Is God a Creationist? The Religious Case against Creation-Science.* New York, 1983.
Gish, Duane T. *Evolution: The Challenge of the Fossil Record.* El Cajon, Calif., 1985.
Godfrey, Laurie R. *Scientists Confront Creationism.* New York, 1983.
Gould, Stephen Jay. *Rocks of Ages: Science and Religion in the Fullness of Life.* New York, 1999.
Hughes, Liz Rank. *Reviews of Creationist Books.* Berkeley, Calif., 1992.
Johnson, Phillip. *Darwin on Trial.* Washington, D.C., 1991.
Kitcher, Philip. *Abusing Science: The Case against Creationism.* Cambridge, Mass., 1982.
Larson, Edward J. *Summer for the Gods.* New York, 1997.
———. *Trial and Error: The American Controversy over Evolution.* New York, 1985.
Matsumura, Molleen. *Voices for Evolution.* Berkeley, Calif., 1995.
McIver, Tom. *Anti-Evolution: An Annotated Bibliography.* Jefferson, N.C., 1988.
———, *Anti-Evolution: A Reader's Guide to Literature Before and After Darwin.* Baltimore, 1992.
Moore, James R. *The Post-Darwinian Controversies.* Cambridge, UK, 1979.
Montagu, Ashley. *Science and Creationism.* New York, 1984.
Morris, Henry M. *Scientific Creationism.* Green Forest, Ariz., 1985.
Nelkin, Dorothy. *The Creation Controversy: Science or Scripture in the Schools.* New York, 1982.
Pennock, Robert. *Tower of Babel: The Evidence against the New Creationism.* Cambridge, 1999.
Ruse, Michael. *Darwinism Defended: A Guide to the Evolution Controversies,* Reading, Mass., 1982.
———, ed. *But Is It Science: The Philosophical Question in the Creation/Evolution Controversy.* Amherst, N.Y., 1988.
Scott, Eugenie C. "Antievolutionism, Scientific Creationism, and Physical Anthropology." *Yearbook of Physical Anthropology* 30 (1987): 21–39.
———. "Antievolutionism and Creationism in the United States." *Annual Review of Anthropology* 26 (1997): 263–89.
White, Andrew Dickson. *A History of the Warfare of Science with Theology in Christendom.* New York, 1896.

DARWIN AND THE LITERARY MIND

Adams, James E. "Woman Red in Tooth and Claw: Nature and the Feminine in Tennyson and Darwin." *Victorian Studies* 33 (1989): 7–27.
Appleman, Philip. *Darwin's Ark,* Bloomington, Ind., 1984; in *New and Selected Poems, 1956–1996.* Fayetteville, Ark., 1996.
———. *Apes and Angels.* New York, 1989.
———, William A. Madden, and Michael Wolff, eds. *1859: Entering an Age of Crisis.* Bloomington, Ind., 1959.
Barzun, Jacques. *Darwin, Marx, Wagner: Critique of a Heritage.* New York, 1941.
Beach, Joseph Warren. *The Concept of Nature in Nineteenth-Century English Poetry.* New York, 1936.
Becker, Anne. *The Transmutation Notebooks: Poems in the Voices of Charles and Emma Darwin.* Washington, D.C., 1996.
Beer, Gillian. *Darwin's Plots: Evolutionary Narrative in Darwin, George Eliot, and Nineteenth-Century Fiction.* London, 1983; Cambridge, UK, 2000.
Blinderman, Charles S. "Huxley, Pater, and Protoplasm." *Journal of the History of Ideas* 42 (1982): 477–86.
Brantlinger, Patrick, ed. *Energy and Entropy: Science and Culture in Victorian Britain.* Bloomington, Ind., 1989.
Carroll, Joseph. *Evolution and Literary Theory.* Columbia, Mo., 1995.
Chapple, J. A. V. *Science and Literature in the Nineteenth Century.* London, 1996.
Culler, A. Dwight. "The Darwinian Revolution and Literary Form." In George Levine and William A. Madden, eds. *The Art of Victorian Prose.* New York, 1968.
Faggen, Robert. *Robert Frost and the Challenge of Darwin.* Ann Arbor, Mich., 1997.
Hyman, Stanley Edgar. *The Tangled Bank: Darwin, Marx, Frazer and Freud as Imaginative Writers.* New York, 1962.

Krasner, James. "A Chaos of Delight: Perception and Illusion in Darwin's Scientific Writing." *Representations* 31 (1990): 118–41.
Krutch, Joseph Wood. *The Modern Temper*. New York, 1929.
Levine, George. *Darwin and the Novelists: Patterns of Science in Victorian Fiction*. Cambridge, Mass., 1988.
——, ed. *One Culture: Essays in Science and Literature*. Madison, Wis., 1987.
——. "Darwin among the Critics." *Victorian Studies* 30 (1987): 253–60.
——. "Darwin and Pain: Why Science Made Shakespeare Nauseating." *Raritan Quarterly* 15 (fall 1995): 97–144.
Morton, Peter. *The Vital Science: Biology and the Literary Imagination, 1860–1900*. London, 1984.
Muller, Herbert J. *The Spirit of Tragedy*. New York, 1956.
Stevens, L. Robert. "Darwin's Humane Reading: The Anaesthetic Man Reconsidered." *Victorian Studies* 26 (1982): 51–63.
Stevenson, Lionel. *Darwin among the Poets*. Chicago, 1932.
Young, Robert. *Darwin's Metaphor: Nature's Place in Victorian Culture*. Cambridge, UK, 1985.

BIBLIOGRAPHIES

Eisen, Sydney, and Bernard V. Lightman. *Victorian Science and Religion: A Bibliography with Emphasis on Evolution, Belief, and Unbelief*. Hampden, Conn., 1984.
Freeman, R. B. *Charles Darwin: A Companion*. Folkestone, 1978.
Ruse, Michael. "The Darwin Industry: A Guide." *Victorian Studies* 39 (1996): 217–35.
Also consult the annual "Victorian Bibliography" in the summer issue of *Victorian Studies* and the annual MLA "International Bibliography" in *PMLA*.
A number of Darwin Web sites exist, some of which include bibliographical resources.

Index